Resistor Theory and Technology

Felix Zandman
Chairman, Scientific Director, and CEO, Vishay Intertechnology, Inc.

Paul-René Simon
Consultant

Joseph Szwarc
Chief Engineer, Vishay Israel Ltd

Distributed by SciTech Publishing, Inc.

Published by Vishay Intertechnology, Inc., Malvern, PA 19355-2120

Produced for Vishay Intertechnology by SciTech Publishing, Inc., Mendham, NJ 07945

This book is exclusively distributed by:

SciTech Publishing, Inc.
c/o William Andrew Inc.
169 Kinderkamack Rd.
Park Ridge, NJ 07656
Phone: (201) 505-4955
Fax: (201) 505-4965
www.williamandrew.com

Printed in the United States of America

Contents

Preface

Today's exciting new products—including cell phones, laptop computers and personal data assistants—are changing the way we live. These products are known for their portability, ease of use, small size and ever-increasing performance. Every one of them uses resistors in one form or another, whether it is a microprocessor with millions of resistors imbedded per square inch, or single resistors designed to sense when the battery is running down.

It is important that engineers know and understand the technology and the science of resistors and resistor materials in order to develop the best overall product designs. With many different capabilities and unique performance characteristics available, engineers can use resistors to address many of their design challenges: power handling, current sensing, ultra high stability, low thermal deviation, thermal sensing, or pulse handling. By matching the right resistor technology to the design requirements, the engineer can optimize the overall product.

Resistor Theory and Technology is intended to fill this need. It addresses the importance of resistive components and other related products. Its foundation is based on Ohm's law, one of the most frequently used relationships in developing active and passive components. Readers will learn about and gain an understanding of why this law and resistors provide the only possibility of measuring or recording a large number of phenomena and their changes. The book clearly and systematically demonstrates that the design and fabrication of resistance components require the application of particularly complex physical phenomena. This is an especially important revelation in light of long-held beliefs that this technology was based on the traditional empirical methods generally associated with a "kitchen-recipe" approach.

Divided into two parts, the book goes from theory and philosophy and progresses to quantification and practical application. Each chapter is intended to stand alone, so the reader may select individual chapters on particular topics and gain benefit from this book. "Part I: Theory" begins with a concise discussion of Ohm's law and resistor properties (Chap. 1), laying the groundwork for the remainder of the book. Chapter 2 covers the effects of reversible phenomena associated with Ohm's law and explains how to maximize and/or minimize them for desired resistor performance. Chapter 3 explains the irreversible phenomena that permanently modify the resistivity of solid and thin-film metals and alloys.

In Chap. 4, the three main conductive materials used to manufacture resistors are covered. The reader will come to understand why these materials require close collaboration among metallurgical research laboratories, alloy developers and resistor manufacturers.

Chapter 5 covers the mechanisms and sources of noise phenomena in resistors and the methods used to measure and minimize them. Chapter 6 then discusses the mechanisms of heat transfer and examines how these concepts can be used to ensure that resistors operate as designed.

The second part of the book, "Part II: Technology," focuses on the practical aspects of these components. Using documentation provided by Vishay, Inc. as well as other manufacturers, the reader is provided with comprehensive examination of the criteria for selecting resistive components for use in specific applications.

Chapter 7 introduces the terminology commonly used to describe the characteristics and performance of resistive components. Chapter 8 then examines and introduces the limitations of resistive circuits that may cause deviations or changes in the actual behavior of a resistor from Ohm's law.

Chapter 9 discusses the principal types of wirewound resistors and their use in power electronics, and further discusses their modified use for surface-mount technology.

Chapter 10 further examines the development and fabrication of precision resistors, whose performance is generally two to three orders of magnitude greater than that of conventional resistors. It covers the stringent requirements that led to these new technologies and discusses their relative merits.

Potentiometers and potentiometric position sensors, as well as their advantages in automated control systems, are the topic of Chap. 11.

Chapter 12 covers the phenomena of anistropic magnetoresistance in materials and its applications in ferromagnetic alloys. It also discusses the growing use of this technology for computer hard drives, audio and video recording, and many other consumer products.

In Chap. 13, nonlinear resistors—now one of the largest growing members of the passive component family—are described and analyzed, including each type's material properties, fabrication technologies, and applications.

The book concludes with a number of useful appendices containing information on specific subjects of foil technology, methods for quality assurance, models for lifetime testing, and resistive strain gages.

Resistor Theory and Technology is intended for students, engineers, and technicians who use these components in their work, for research, or for product development.

Introduction

At a time when microprocessors containing millions of transistors, fiber optic networks, HDSL (high-bit-rate digital subscriber line) and ADSL (asymmetrical digital subscriber line)—two new products in the field of telecommunication and fast data transfers—and other advanced technologies of electronic circuits and systems are becoming commonplace, it may appear somewhat odd and anachronistic to write a book on resistive components and related products.

It is worth pointing out, however, that the physical relationship expressed by Ohm's law is still one of the most frequently used relationships in developing active and passive components. Without Ohm's law, and without resistors, there would be no possibility of measuring or recording a large number of phenomena and their changes.

The controls of a Boeing 737 or Airbus 320 are activated by the pilot by means of resistive movement sensors. Although the digital data on the tracks of a computer are read and written with inductive devices, magnetoresistive sensors can be used for the readout of data from the hard disks. Since two or three years ago, this technology is now widely used by the market. Old-fashioned wirewound resistors are still used in many electronic applications, especially power controls for fast railway trains, or shunts applications in power sensors.

At the same time, the constraints of miniaturization imposed by passive components have led manufacturers of resistive products to adopt methods of miniaturization and integration currently used in the production of printed circuits. A 16-bit R-2R network now requires a silicon area of only 1 to 2 mm^2.

Along with these developments, the performance of resistive components, in terms of precision, useful life, and temperature coefficient, has increased by a factor of between 10 and 100 over the past 20 years. As a result of the requirements imposed by military standards (MIL-STD), their reliability is now as good as that of the best components and active integrated circuits.

All of these so-called "high-tech" technologies are involved in the production of resistive products: precision metallurgy for resistive alloys, materials science for thermistors and magnetoresistive products, vacuum deposition, vacuum implantation, and photolithography for integrated resistive components, etc.

Printed circuit boards are largely populated with integrated circuits, passive electronic components (resistors, capacitors, inductors) and discrete semiconductors (diodes and transistors). It is interesting to note that in recent years the growth of passive components has been increasing faster than that of integrated circuits. In other words, as integrated circuits become more and more sophisticated, the number of passive components surrounding them is increasing steadily. For example, between 1991 and 1996, the use of integrated circuits increased from 35 billion per year to 42 billion per year, an average annual increase of 5 percent, while the use of tantalum and multilayer ceramic capacitors increased from 115 billion to 210 billion per year, an average annual increase of 17 percent. Chip resistor production also increased in the same period of time from 125 billion to 230 billion per year, also an annual increase of 17 percent. While it was a commonly held belief in the 1960s that discrete components would disappear because of the advent of integrated circuits, in fact, the opposite has occurred, and they have grown faster than ICs. This can be explained by the fact that the real estate on the semiconductor wafer is expensive and therefore mostly used for active devices, which are more costly than passive devices. Furthermore, today, design changes are extremely frequent (with some companies introducing new computers and cellular telephones every few months) and such changes can't be implemented quickly enough because the redesign of integrated circuits is a time-consuming and costly process. Also, certain passive components, such as power devices, precision resistors, and high-CV capacitors, cannot be integrated into a silicon

IC. Consequently, frequent design changes are accomplished by adding passive components and changing their configuration in the circuit.

This work attempts to demonstrate that the design and fabrication of resistive components require the application of particularly complex physical phenomena and are no longer based on the traditional empirical methods generally associated with a "kitchen-recipe" approach.

The book is addressed to students, engineers, and technicians who use these components in their work, for research, or for product development. The book, however, does not deal with the very important issue of resistance in semiconductor devices and the parasitic resistance in capacitors and inductors.

PART I: THEORY

1

Ohm's Law—Resistance

1. INTRODUCTION

A material medium generally contains a considerable number of charge carriers located at microscopic distances from one another, usually on the order of a nanometer or less. However, it is not possible to look upon the properties and trajectories of every carrier. Therefore, using the statistical physical description of material properties, the average distance used to characterize the distribution of the carriers is of macroscopic order. Conductors have an essential role in electronic technology. They have the properties that some charge carriers are free to move to macroscopic distances when exposed to the action of an electric field. The average distances used to characterize the distribution of these carriers are, therefore, macroscopic.

2. ELECTROKINETICS

We can examine (Fig. 1) the case of a volume $\Delta\mathcal{V}$ in which there is a distribution of n charge carriers in which each carrier q_i has mass m_i and velocity $\vec{v}_i$ with respect to the laboratory reference point.

We characterize the transport of matter associated with a carrier by its momentum $m_i\vec{v}_i$. We can also characterize the charge transport by

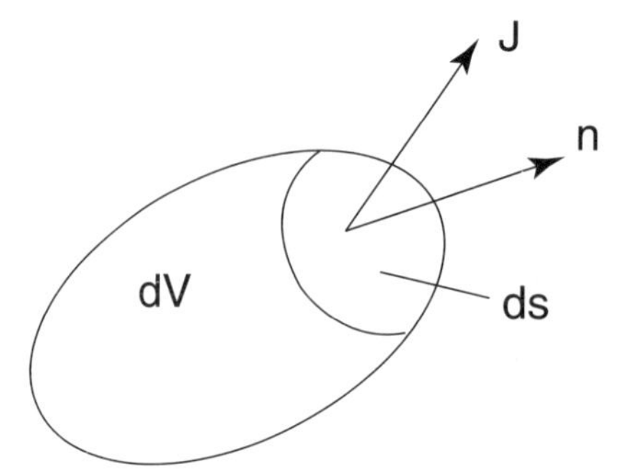

Fig. 1 Charge carriers in a volume dV.

introducing the quantity $q_j\vec{v}_i$. We can then define, at each point P of the charge distribution, the *volume current vector* $\vec{J}$ by:

$$\vec{J} = \frac{1}{\Delta V}\sum_{i=1}^{i=n} q_j\vec{v}_i \tag{1.1}$$

$\vec{J}$ is calculated in several stages. We first separate the different types of carriers, then calculate the average velocity of all carriers of each type.

As a result $\vec{J}$ can be expressed as

$$\vec{J} = \sum^{\alpha} j_\alpha \qquad \text{with} \qquad \vec{J}_\alpha = N_\alpha q_\alpha \vec{v}_\alpha = \rho_\alpha \vec{v}_\alpha \tag{1.2}$$

in which N_α and $N_\alpha q_\alpha = \rho_\alpha$ are, respectively, the density and the volume charge of carriers of type α.

The volume current vector here appears as the vector sum of the partial volume currents. It should be pointed out, however, that among the various types of carriers α, the only ones that participate in the electrical current are those whose drift velocity $\vec{v}_\alpha$ is not zero, whereas all the carriers contribute to the volume charge.[1,2] The condition $\vec{v}_\alpha = \text{Ct}$ represents the isotropy of the velocity distribution of each carrier α and not necessarily their state of rest.

---

[1]In practice we must account for the existence of a velocity distribution law when we sum $\vec{v}_\alpha$. In a material environment this distribution obeys a statistical law that depends on the nature of the particles, their density and temperature.

[2]The reader may wish to consult the following works, which cover the theory of electric fields and conductivity in considerable depth.

Vassallo, C., *Electromagnétique classique dans la matière*. Dunod: 1980.

Quéré, Y., *Physique des matériaux*. Ecole Polytechnique Ellipses: 1988.

Ziman, *Electrons and Phonons*. Oxford Clarendon Press: 1962.

R. Kubo and T. Nagamiya, *Solid State Physics*. McGraw-Hill: 1969.

2.1. Solids

In solid conducting materials we normally distinguish two categories of charges. The first consists of "bound" charges whose displacement is microscopic and does not contribute to the current density in a *steady-state environment.*[3] In the other, charges are "mobile," or "free," and are the only ones that contribute to the current on a macroscopic level. In silver, for example, on average, one electron can move through the lattice of fixed Ag^+ ions. If ρ, ρ_f, and ρ_m represent the total, fixed, and mobile volume charges, it is obvious that:

$$\rho = \rho_f + \rho_m \tag{1.3}$$

whereas if there is only a single type of mobile charge having a drift velocity $\vec{v}_m$:

$$\vec{J} = \rho_m \vec{v}_m \tag{1.4}$$

When the material is in a uniform field, as implied by the above description, the velocity $\vec{v}$ is much lower than the velocities of thermal motion of the charge carriers in equilibrium with the crystal lattice. For example, the mean square velocity $\langle v^2 \rangle$ of a "perfect electron gas" at room temperature is given by:

$$m\langle v^2 \rangle = 3kT \cong 10^{-20}\,J \qquad \text{or} \qquad v = 10^5\ \text{m/s} \tag{1.5}$$

In a metal, in which each atom produces a free electron, there would be 10^{29} electrons/m^3 and a current of 1 A/mm^2, which is typical of currents in electric wires and resistors. This would correspond to an average velocity $v \cong 10^6/10^{29-19} = 10^{-4}$ m/s, far below the velocity of thermal motion. This is not always the case, especially in the majority of high-performance electronic circuits, which use very high current densities.

2.2 Magnitude of Electric Current

The magnitude of the electric current crossing a surface with orientation S is defined by the flux of the vector $\vec{J}$ across this surface (Fig. 1).

$$I = \iint_s \vec{J} \cdot \vec{n}\, dS \tag{1.6}$$

[3]Under certain nonsteady-state conditions, these charges can participate in conductivity and thus introduce defects in resistors.

For an element of surface dS about a point P in a material medium, the corresponding elementary current is written as:

$$dI = \vec{J} \cdot \vec{n} dS \tag{1.7}$$

If the distribution contains only one type of mobile charge, then

$$dI = \rho_m \vec{v}_m \cdot \vec{n} dS \tag{1.8}$$

The current magnitude is thus an algebraic value. It is positive if the positive direction arbitrarily chosen for $\vec{n}$ coincides with the direction of movement of the positive charges.

The SI unit of current being the ampere (A), the current volume is expressed in A/m^2 (in practice A/cm^2 is used). The norm of $\vec{J}$ represents the current that crosses the unit of surface perpendicular to $\vec{J}$.

$$J = \|\vec{J}\| = dI/d\Sigma \tag{1.9}$$

in which $d\Sigma = dS \cos\theta$ and $\theta = (\vec{n} \cdot \vec{J})$.

2.3 Charge Conservation

We can easily demonstrate[4] that there is a local conservation of electric charge according to the relation:

$$\frac{\partial \rho}{\partial t} + \text{div}\, \vec{J} = 0 \tag{1.10}$$

When there are carriers of several types of α, as in an electrolyte, for example, these can neutralize one another through recombination. Inversely, charges of opposite sign can be created during a process of regeneration (anion–cation pairs). The conservation of charge for a given type is, therefore, not respected. In this case we can write:

$$\frac{\partial \rho_\alpha}{\partial t} = \text{div}(-\vec{J}_\alpha) + p_\alpha \tag{1.11}$$

Here p_α is the rate of creation ($p_\alpha > 0$) or elimination ($p_\alpha < 0$) of α carriers. The overall rate for all types of carriers leads naturally to Eq. 1.9.

[4]Vassallo, *op. cit.*

3. STEADY-STATE OHM'S LAW

A material is a *conductor* if it contains charge carriers capable of moving on a macroscopic scale. In an insulating medium, however, all the charges are bound, which reduces the extent of motion to a microscopic value.[5]

Ohm's law, which was developed by the physicist G. S. Ohm, is a relation between the electrical currents that appear in a given conductor and the external factors that produce them. As such, the law does not possess the universality that characterizes the laws of electromagnetism. We refer to it as a *constituent law* or a *local relation.*

3.1 Conductivity

The appearance of an electric current in a conducting material represents the disruption of a state of equilibrium. Experimentally, we find that a current can appear in a conductor if there is a gradient of temperature, electrical potential, or carrier concentration. In this chapter we will only examine the case of a current caused by a potential gradient; that is, by a macroscopic field $\vec{E}$, the concentration of carriers and the temperature being considered as constant.

The relation between the volume vector $\vec{J}$ and the field $\vec{E}$ depends on the environment under consideration and the value of $\vec{E}$. Experimental evidence shows that *if the deviation with respect to equilibrium is sufficiently small* (the norm of $\vec{E}$ being less than $1.5–2.4 \cdot 10^6$ V/m), the vector $\vec{J}$ is, as a first approximation, proportional to the field $\vec{E}$ that caused it. For any point in the interior of the material, we can write:

$$\vec{J} = \sigma_0 \vec{E} \tag{1.12}$$

This linear relation is known as the *local Ohm's law.* The coefficient σ_0 is called the *conductivity* of the medium. When this relation is verified, the medium is said to be *linear* for the phenomenon of conduction. Therefore, σ_0 does not depend on $\vec{E}$. It is also *uniform* if the medium is homogeneous. In the majority of materials $\vec{J}$ is parallel to $\vec{E}$, and the medium is *isotropic* for this property. Conductivity is then represented by a positive real scalar.

[5]These microscopic movements can, however, induce polarization and magnetization currents. For the moment we will simply point out that these do not occur in a steady-state environment in a nonmagnetic conductor.

The standard unit for conductivity is siemens per meter (S/m), named after the engineer W. Siemens. In electronics and electrical engineering we introduce the concept of inverse conductivity, known as the *resistivity* of the medium and generally represented as $\rho_0 = 1/\sigma_0$ (this should not be confused with the volume charge).

In certain materials $\vec{J}$ is not parallel to $\vec{E}$ and the conductivity is then a tensor that is specified in a given base in a square matrix of nine elements.

$$\sigma_0 = [\sigma_{0ij}](i, j = 1, 2, 3) \tag{1.13}$$

In such an *anisotropic* medium, Ohm's law is written:

$$\vec{J}_1 = \sum_{j=1}^{j} \sigma_{0ij} \vec{E}_j \tag{1.14}$$

Anisotropy is directly related to atomic structure. A cubic monocrystalline conductor, therefore, possesses isotropic conductivity. A material such as graphite, however, which consists of a series of layers of material, is anisotropic. Its conductivity is greater in a direction parallel to the plane of the layers than perpendicular to them. The materials most frequently used are polycrystalline or amorphous, and, therefore, are isotropes on the macroscopic level.

3.2 Electrostatic Equilibrium of a Conductor

The absence of a current vector ($\vec{J} = 0$) in a conductor of uniform temperature and composition defines a state of *electrostatic equilibrium.* Such an equilibrium state corresponds to the absence of a macroscopic electrostatic field at any point within the conductor, $\vec{E}_j = 0$. The electrostatic potential is therefore uniform: $V_{int} =$ constant. In a conductor, the environment relation determines the value of the field *without the use of electrostatics theorems.*

3.3 Conductors in a Steady-State Environment

If the conductor through which a steady-state current flows is homogeneous, the continuity equation and Ohm's law imply that:

$$\text{div}(\vec{J}) = \text{div}(\sigma_0 \vec{E}) = \sigma_0 \cdot \text{div}(\vec{E}) = 0 \tag{1.15}$$

As a result, the field $\vec{E}$ in such a conductor satisfies the following local relations:

$$\operatorname{div}(\vec{E}_{in}) = 0 \quad \text{and} \quad \vec{E} = -\vec{\text{grad}}(V) \tag{1.16}$$

in which $\vec{E}_{in}$ is the component of the field perpendicular to the surface and V is the potential difference between the two surfaces perpendicular to the conductor.

It should be noted that the first equation does not necessarily imply that $\vec{E}_{in}$ is uniform. For example, in a wire of variable cross-section, the current I is the same throughout any cross-section. As a result $\vec{J}$ and therefore, $\vec{E}$, the component of the field perpendicular to $\vec{E}_{in}$, vary from one cross-section to the next according to the following equations:

$$I = (\vec{J}_1 \cdot \vec{n}_1)S_1 = (\vec{J}_2 \cdot \vec{n}_2)S_2 \tag{1.17}$$

$$\sigma_0(\vec{E}_1 \cdot \vec{n}_1)S_1 = \sigma_0(\vec{E}_2 \cdot \vec{n}_2)S_2 \tag{1.18}$$

and

$$(\vec{E}_2 \cdot \vec{n}_2)/(\vec{E}_1 \cdot \vec{n}_1) = S_1/S_2 \tag{1.19}$$

Based on the equation $\operatorname{div}(\vec{E}) = \rho/\varepsilon_0$, which is the local representation of Gauss's theorem within a conductor, we can deduce, since $\operatorname{div}(\vec{E}) = 0$ inside the conductor, that the average volume charge ρ is zero at every point. Consequently, the density of the mobile charges is compensated by that of the bound charges and an excess charge can occur only at the surface of the conductor.

The volume neutrality imposed by Ohm's law enables us to state that in a steady-state environment a nonneutral medium does not satisfy Ohm's law. This is the case for resistors subject to high electrical fields, which induce diffusion phenomena, or that of semiconductors. The relation $\vec{J} = \sigma_0 \vec{E}$ is, in this case, not confirmed.

4. INTEGRAL FORM OF OHM'S LAW

Consider a segment of an isotropic conductor lying between two sections S_1 and S_2, between which a steady-state potential difference has been established: $U = V_1 - V_2$. Because the same current I passes through every cross-section of this conductor, the voltage U and the current I can be calculated from local values of $\vec{J}$ and $\vec{E}$, since:

$$U = \int_{M_1}^{M_2} \vec{E} \cdot d\vec{r} \qquad \text{and} \qquad I = \iint_S \vec{J} \cdot \vec{n}\, dS \tag{1.20}$$

in which M_1 and M_2 are the surfaces of the conductor to which the voltage is applied, $\vec{J}$ and $\vec{E}$ being connected at every point by the local expression of Ohm's law $\vec{J} = \sigma_0 \vec{E}$.

The relation $R = U/I$ is unchanged if the electric field (and therefore the current) is multiplied by a constant. Moreover, since $[\vec{n}]$ and $[d_{\vec{r}}]$ are oriented in the same direction (σ_0 is positive), R is also positive. We refer to this as the resistance of that part of the conductor under consideration. It depends on the type, geometry, and relationship of the conductor to the electric field. The legal unit of resistance is the Ohm (Ω).[6]

The calculation of the resistance of a homogeneous conductor is based on the consideration of a tubular current element with an orthogonal cross-section dS and length dl. Because this tubular element is also a tube of force, we can write:

$$dV = -\vec{E} \cdot d\vec{r} = -\vec{E} \cdot d\vec{l} \quad \text{and} \quad dI = \vec{J} \cdot \vec{n} dS = J_n dS = \sigma_0 E_n dS \tag{1.21}$$

If the conductor is homogeneous and of constant cross-section S between the two equipotential surfaces separated by distance L, the electric field is uniform. Consequently:

$$U = -\int_L dV = EL \quad \text{and} \quad I = \iint_S dI = \sigma_0 ES \tag{1.22}$$

Based on this we can express the resistance of a conductor as:

$$\boxed{R = \rho_0 \frac{L}{S}} \tag{1.23}$$

R dimension in the SI system is $L^2T^{-1}MQ^{-2}$ (ρ_0 dimension being $L^3T^{-1}MQ^{-2}$), which is expressed in ohm.m or in usual units, ohm.cm.

5. ELEMENTARY THEORY OF CONDUCTIVITY (P. K. Drude Model)

In a conductor, the mobile charges are not completely free since they interact among themselves and with the fixed charges in the material, whereas

[6]The recent discovery of the quantized Hall effect enables us to express a new legal definition of the Ohm based on the fundamental constants $q_e = 1.6 \times 10^{-19}$ and $h = 6.63 \times 10^{-34}$ joule-second. Because the ratio h/q^2 has the dimensions of resistance, we can use its value as a standard: $h/q^2 = 25812.8056\ \Omega$.

in an electrolyte the mobile particles are ions that move among neutral molecules. Interactions are then described as collisions among the different particles. Interactions in a solid cannot be described as simply, however, and must be formulated within the framework of quantum mechanics.

However, a phenomenological approach that can be used to calculate the orders of magnitude of σ_0 was proposed by P. K. Drude[7] in 1902. It consists of representing the action of a material medium on mobile charges of density ρ_m, contained in an element of volume dV, through the force of viscous friction. For a given type of charge density ρ_{me}, this force, which acts against the drift velocity $\vec{v}$, can be expressed:

$$d\vec{F}_m = -\alpha\vec{v}dV \qquad \text{with} \qquad \alpha > 0 \tag{1.24}$$

Assuming the movement of charges is caused by a macroscopic electric field $\vec{E}$ within the medium, the fundamental law of mechanics, applied to the element of volume dV, gives:

$$d\vec{F}_e + d\vec{F}_m = \rho_{me}\frac{d\vec{v}}{dt}dV \tag{1.25}$$

where $d\vec{F}_e = \rho_m\vec{E}dV$ is the electrical force.

If m and q represent the effective mass and charge of the carriers under consideration, and if $\mathfrak{R}$ represents their density, then:

$$\rho_{me} = \mathfrak{R}_m \qquad \text{and} \qquad \rho_m = \mathfrak{R}_q \tag{1.26}$$

This results in the following differential equation:

$$\frac{d\vec{v}}{dt} = \frac{q\vec{E}}{m} - \frac{\alpha}{\mathfrak{R}m}\vec{v} \tag{1.27}$$

which also can be written as:

$$\frac{d\vec{v}}{dt} + \frac{\vec{v}}{\tau} = \frac{q\vec{E}}{m} \tag{1.28}$$

where $\tau = \mathfrak{R}m/\alpha$. The physical significance of τ, which is uniform for a given time period, appears clearly if we study the change in velocity $\vec{v}$ over time when the electric field is suddenly cut off at a given time taken as the point of origin. Under these conditions, we have:

$$\frac{d\vec{v}}{dt} + \frac{\vec{v}}{\tau} = 0 \qquad \text{for} \qquad t \geq 0 \tag{1.29}$$

[7]Drude, P. K., *Ann. d. Physik,* 7 (1902): 687.

The solution to this first-order differential equation can be expressed as $\vec{v} = \vec{v}_0 \cdot \exp(-t/\tau)$, in which $\vec{v}_0$ is the velocity at the point of origin. In this case τ represents the exponential rate of decrease in the drift velocity[8] $\vec{v}$. After several τ this velocity is practically zero and the distribution of velocities becomes isotropic. Since the memory of collective movement is lost in this situation, we call τ the *relaxation time*. Its value is directly related to the frequency of collision: As the frequency increases, the energy that is dissipated increases and τ decreases.

The solution to the differential equation with the second member can be obtained by adding to the previous solution a specific solution of the global differential equation that defines the established environment.

By assuming $\vec{v} =$ constant, which implies that $d\vec{v}/dt$ is zero, we obtain, through Eq. 1.28, the relation

$$\vec{v} = \frac{q\tau}{m}\vec{E} = \mu\vec{E}$$

The mean drift velocity is, therefore, proportional to the electric field. The coefficient μ, called *mobility* (in m^2/volt-sec), characterizes a specific type of carrier within the medium. When there are several types of carriers simultaneously, the volume current is summed for the different types of carriers.

The resulting volume current can be expressed as:

$$\vec{J} = \sum^{\alpha} N_\alpha q_\alpha \vec{v}_\alpha = \sigma_0 \vec{E} \tag{1.30}$$

Ohm's law such established is based on the limitation of the drift velocity of the carriers resulting from their interactions with the material medium.

[8]The relaxation time is defined by an exponential statistical distribution such that the probability *P(t)* that a carrier had not undergone a collision during time *t* and will undergo a collision between *t* and $t + dt$ is:

$$P(t) = \frac{1}{\tau} \cdot e^{-t/\tau}$$

As a result, the time τ is simply the *average time of flight* of the carriers between collisions. We can also calculate the mean quadratic value:

$$\langle t^2 \rangle = \int_0^\infty t^2 P(t)dt = 2\tau^2.$$

In a metal or metallic alloy, an electric current is created by the ensemble movement of the conduction electrons (carriers which are not bound to the constituent atoms of the material) with volume density $\mathfrak{N}$. As a result, for μ and σ_0 we have:

$$\boxed{\mu = \frac{q_e \tau}{m} \qquad \text{and} \qquad \sigma_0 = \frac{\mathfrak{N} q_e^2 \tau}{m}} \tag{1.31}$$[9]

For the majority of metals and metallic alloys, the order of magnitude of σ_0 is $0.12 - 6 \times 10^7\ \text{ohm}^{-1} \cdot \text{m}^{-1}$.

Based on the measured values of σ_0 and $\mathfrak{N}$ ($\mathfrak{N} = 8 \times 10^{28}\ \text{m}^{-3}$), it is possible to determine the order of magnitude of τ, which is $3 \times 10^{-14}\ s$ or 0.03 ps.

This small value for τ, due to the low effective mass (on the order of 1.9 times the mass of the electron at rest), shows that the electrons react very quickly to electrical action. The transition state is thus negligible as long as the typical durations for the change in electrical fields are considerably higher than 10^{-14} s (10,000 GHz).

In semiconductors, the value of σ_0 is between 10^{-5} and 10^4 S/m. However, the true distinction between a metal or metallic alloy, a semiconductor, or an insulator is to be found in the mechanisms by which conduction occurs, for these enable us to explain the way conductivity varies with changes in temperature, the applied fields, or the introduction of impurities.

5.1 Conduction in a Sinusoidal Environment

In a sinusoidal environment (time dependence on the variables in $e^{-i\omega t}$) the equation of motion in Eq. 1.27 becomes (by assuming that $\vec{E}$ is parallel to the x axis):

[9] A detailed study of conductivity shows that Drude's relation is only applicable for an isotropic monovalent metal with a simple conduction band. (In the region of space occupied by the solid, the presence of N positive ions creates an attraction potential for the electrons. But aside from this global action, their local activity is ignored.) However, its application to metals of more complex electronic structure enables us to provide a semiquantitative description of conduction phenomena in these metals. In this case, however, the viscous force must take into account the band structures in the crystal, which implies a quantum approach to the theory. Contrary to Drude's hypothesis, according to which the electrons should obey a Maxwell–Boltzmann statistical distribution, they obey Fermi–Dirac statistics. Using this approach we can apply the Pauli exclusion principle to the electron gas and thus explain the particular behavior of conductivity in the transition metals and their alloys, which, as we shall show, are among the principal materials used to manufacture resistors. Readers who are interested in these issues should refer to Y. Quéré, *Physique des Matériaux,* Ellipse, 1989.

$$-mi\omega\vec{v}_x = q\vec{E} - \frac{m\vec{v}_x}{\tau} \tag{1.32}$$[10]

or

$$\vec{v}_x = \frac{q\vec{E}\tau}{m} \cdot \frac{1}{1 - i\omega\tau} \tag{1.33}$$

The current density:

$$\vec{J}_x = \frac{\Re q^2\tau}{m} \cdot \frac{1}{1 - i\omega\tau} \cdot \vec{E} \tag{1.34}$$

leads to a complex equation for conductivity:

$$\sigma = \frac{\sigma_0}{1 - i\omega\tau} \tag{1.35}$$

If $\omega\tau \gg 1$, then conductivity can be reduced to the imaginary term:

$$\sigma = \frac{i\sigma_0}{\omega\tau} = \frac{i\Re q^2}{m\omega^2} \tag{1.36}$$

The fact that the conductivity is an imaginary number signifies that $\vec{J}$ is in quadrature with $\vec{E}$ and consequently that the power dissipated by the Joule effect is zero.

In a polarizable medium, where there are charges (volume density ρ) and currents (current density $\vec{J}$), Maxwell's equations[11] can be written:

$$\text{div}\vec{D} = \rho, \quad \text{div}\vec{B} = 0, \quad \text{rot}\,\vec{B} = \mu_0\vec{J} + \varepsilon_0\mu_0\frac{\partial\vec{E}}{\partial t}, \quad \text{rot}\,\vec{E} = -\frac{\partial\vec{B}}{\partial t} \tag{1.37}$$

Applying these equations to a monochromatic plane wave propagating through a conductor and whose spatiotemporal frequency is $\exp i(\vec{k}\cdot\vec{r} - \omega t)$, we can write:

$$i\vec{k} \wedge \vec{B} = \mu_0\vec{J} - i\varepsilon_0\mu_0\omega\vec{E} \tag{1.38}$$

[10]In writing this equation we have ignored the Laplacian force in the presence of the electrostatic force. In effect, for an electromagnetic wave $(\vec{E}, \vec{B})$, we have $|\vec{E}|/|\vec{B}| = c$, the speed of light. The relationship between the two previous forces is therefore v/c, which is negligible.

[11]Jackson, J. D., *Classical Electrodynamics.* New York: John Wiley & Sons, 1962, p. 223.

in which μ_0 and ε_0 are respectively the magnetic permeability and dielectric constant of a vacuum.

Ohm's law enables us to rewrite this equation as follows:

$$i\vec{k} \wedge \vec{B} = i\varepsilon\mu_0\omega\vec{E} \tag{1.39}$$

by introducing the complex dielectric constant:

$$\varepsilon = \varepsilon_0 + \frac{i\sigma}{\omega} = \varepsilon_0 + \frac{\Re q^2}{m} \cdot \frac{1}{\omega^2 + i\omega/\tau} \tag{1.40}$$

This expression represents the complex dielectric constant of a conductive medium carrying charges q. In the case of metals and alloys, these charges are electrons.

5.2 Skin Effect

In the case when $\omega\tau \ll 1$, the dielectric constant takes the following form:

$$\varepsilon = \varepsilon_0 + i\frac{\Re q^2 \tau}{m\omega} = \varepsilon_0\left[1 + i\frac{\omega_p^2\tau}{\omega}\right]$$

where $\omega_p = \Re q^2/m\varepsilon_0$ is the plasma frequency.[12] In this case the dispersion equation for a monochromatic plane wave $k^2 = \mu_0\varepsilon\omega^2$ leads to an expression in which k is imaginary:

$$k = (\mu_0\sigma_0\omega)^{1/2} \cdot \frac{1 + i}{\sqrt{2}} = k' + ik'' \tag{1.41}$$

and the wavelength assumes the form $e^{i(kz-\omega t)} = e^{i(k'z-\omega t)}e^{-k''z}$. The loss associated with the imaginary term appearing in k can be characterized by the depth of penetration (or skin depth). This is the distance by which

[12]The plasma frequency or plasmon corresponds to the total oscillation of an electron gas, resulting in localized charge densities. Given the condition $\text{div}(\vec{E}) = 0$, a global return force $-q\vec{E}$ affecting all the electrons responsible for the charge density is created, such that if λ is the characteristic dimension of the charge density, then we can write $m\frac{d^2\lambda}{dt^2} = -\Re q^2\lambda$, based on which $\lambda = \lambda_0\cos(\omega_p + \phi)$. These oscillations, called *plasma oscillations,* can be observed by recording the energy losses in a beam of electrons crossing a thin layer. This measurement reveals a discrete structure, the losses being equal to an integer multiple of the energy associated with the plasmon. For example, the elementary plasmon possesses an energy of 14 eV. Depending on whether or not the signals are such that $\omega\tau \gg 1$, the metals will be opaque to wavelengths greater than $\lambda_p = c/v_p$, and transparent in the opposite case.

TABLE 1

Metal	Conductivity ($ohm^{-1}m^{-1}$)	Skin Depth (mm) 50 Hz	1 kHz	1 Mhz
Copper	$5.8 \cdot 10^7$	9.3	2.1	0.066
Aluminum	$3.5 \cdot 10^7$	12.1	2.7	0.065
Gold	$4 \cdot 10^7$	10.6	2.38	0.075
Silver	$6 \cdot 10^7$	9.1	2.03	0.064

the field inside the conductor is reduced by the ratio $1/\varepsilon$; that is, $\delta = 1/k' = \sqrt{2/\mu_0\sigma\omega}$. The electromagnetic wave is thus localized in a layer of thickness δ. This is called the *skin effect.*

In the presence of a pulse ω, a cylindrical conductor will present a resistance of R_ω to the passage of a current greater than the resistance R_0 in a continuous medium, that is $R \cong L/2\sigma\pi r\delta$ rather than $R_0 = L/\sigma\pi r^2$ for a resistance of length, L, and radius, r.

6. JOULE'S LAW

The assumption of friction forces associated with the charge carriers implies that collisions occur with the fixed ions in the lattice. *These collisions must not favor any particular direction.* In other words the carrier must "leave the collision" with a velocity that has a random direction. For this reason the velocity $\vec{v}'$ of the carriers as they leave the collisions does not contribute to the diffusion velocity $\vec{v}$.

This implies that the kinetic energy $1/2\, mv^2$ is lost by the carrier during the collision and transferred to the solid. This is known as the *Joule effect*. For a given time of flight before collision t, this energy is $\Delta W = (qEt)^2/2m$. The mean value of this energy is obviously given by:

$$\langle \Delta W \rangle = \frac{1}{m}(qE\tau)^2 \tag{1.42}$$

The number of collisions per unit volume is $\Re/\tau$ and the total energy transferred per unit of time and volume is $w = \Re(qE)^2 m\tau$, from which we derive:

$$w = \frac{J^2}{\sigma_0} \quad \text{(Joule's law)} \tag{1.43}$$

If the material consists of a cylinder of length, L, and cross-section, s, subject to a field $\vec{E}$, parallel to its length, then it is subject to a current $I = (\vec{J} \cdot \vec{n})s$ and the energy given off per unit of time is:

$$W = Lsw \qquad \text{or} \qquad W = RI^2 \tag{1.44}$$

7. THE LIMITS OF VALIDITY OF OHM'S LAW

Ohm's law applied locally assumes that conductivity is independent of the value of the applied electric field $\vec{E}$. This law is no longer valid when the relaxation times τ or carrier densities $\mathfrak{N}$ depend on $\vec{E}$.

(a) *Deviations associated with the relaxation time.* Relaxation of the distribution of the individual velocities of the carriers is only possible if the perturbation from the electric field $\vec{E}$ is weak. The constant τ is the average duration between two successive collisions; the average distance traveled during this time being the *mean free path between collisions* λ. This value is related to the mean velocity $\langle v \rangle$ by the relation $\lambda = \langle v \rangle \tau$. The mean velocity $\langle v \rangle$ is independent of $\vec{E}$ only if the field causes a negligible change in the mean kinetic energy ε_M of the carriers. In the case of a metal, $\langle v \rangle$ is very near the velocity of the electrons located near the Fermi surface; that is, $v_F = \langle v \rangle \approx 1.6 \times 10^6$ m/s, then $\lambda = 3 \times 10^{-8}$ m. By assuming that the field is constant between two collisions, we obtain $\Delta\varepsilon_M = qE\lambda$. The condition of validity for Ohm's law then becomes:

$$qE\lambda \ll \varepsilon_M + \frac{1}{2} m_F^2 \tag{1.45}$$

Based on which

$$E \ll \frac{\varepsilon_M}{q\lambda} \tag{1.46}$$

or $E \ll 2.4 \times 10^8$ V/m. It is possible, however, that the local values of the field near crystal defects or grain boundaries are close to this limit. As we shall see, this can lead to significant deviations in conductivity compared with Ohm's law.

(b) *Deviations associated with density.* Density can vary, especially if the kinetic energy gain from the electric field exceeds the creation threshold of an ionization pair. This can be sufficient to trigger so-called "avalanche" processes. In such cases current intensity increases very rapidly.

(c) *Nonsteady-state conditions.* Charge carriers react to an external stimulus only after a certain value of τ. If the electric field varies significantly during a time T that is less than τ, Ohm's law is no longer respected. Note that this limit does not impose severe restrictions on the majority of conducting materials.

8. ELECTRODYNAMICS OF STEADY-STATE ENVIRONMENTS

The action of an electromagnetic field $\vec{E}$, $\vec{B}$ that is stationary with respect to a charge q with a velocity $\vec{v}$ with respect to a fixed reference, is fully described by Lorentz's law:

$$\vec{F} = q(\vec{E} + \vec{v} \wedge \vec{B}) \tag{1.47}$$

In a conducting medium all the mobile charges contained in an element of volume dV are subject to this force:

$$d\vec{F}_m = \rho_m(\vec{E} + \vec{v} \wedge \vec{B})d\mathcal{V} \tag{1.48}$$

as well as to the interaction of the fixed charges referred to as frictional forces:

$$d\vec{F}_{fm} = -\alpha\vec{v}d\mathcal{V} \tag{1.49}$$

Because the force resultant is zero in a steady-state environment, we can determine the drift velocity as follows:

$$\vec{v} = \mu(\vec{E} + \vec{v} \wedge \vec{B})$$

and Ohm's law can then be written:

$$\vec{J} = \sigma_0\left(\vec{E} + \frac{\vec{J}}{\Re q} \wedge \vec{B}\right) \tag{1.50}$$

in which, as above, $\sigma_0 = \Re q^2\tau/m$.

In the presence of a magnetic field, therefore, the volume current is no longer colinear with the electric field, even in an isotropic environment. Eq. 1.50 can be expressed as:

$$\vec{J} = [\sigma_0(\vec{B})]\vec{E} \tag{1.51}$$

where $[\sigma_0(\vec{B})]$ is the conductivity tensor described by a 3×3 matrix.

The electric field thus contains two elements, the first $\vec{E}_{//} = \vec{J}/\sigma_0$ being parallel to $\vec{J}$ and the second $\vec{E}_\perp = -\vec{J} \wedge \vec{B}/(\Re q)$ being perpendicular to

$\vec{J}$. The ratio of these two terms reaches a maximum if $\vec{J}$ is orthogonal to $\vec{B}$ and is equal to $|\mu|\vec{B}$. We see that it is directly related to the mobility of the carriers. This effect remains very weak for the majority of conducting materials, but it can't be ignored for semiconductor materials and materials of very low conductivity intended for use in high-ohmic value resistors.

The drift of a resistance that is orthogonal to a field of 1 Tesla in a material with a mobility of $4 \cdot 10^{-2}$ SI (cermet) is on the order of 1,600 ppm.

9. CONCLUSION

Ohm's law is only a linear approximation that can be applied to conductive phenomena (resistivity) in crystals. It can be used for all conducting or semiconducting materials, and enables us to calculate the characteristics of resistive components produced from these materials. Depending on the conditions of use or the signals employed, second-order effects will occur and a practitioner skilled in the art will have to measure these effects and take them into account when developing components. In the chapters following we will examine these effects and, on the basis of external constraints, calculate their orders of magnitude in such a way as to specify the conditions under which the validity of Ohm's law might be called into question.

2

Reversible Phenomena Associated with Ohm's Law

1. INTRODUCTION

The value of the conductivity σ depends on the physical properties of the material and can be considered as a characteristic constant for each material, when external and internal conditions remain at rest. However, when not too large external constraints are applied to the materials, reversible variations result in the σ's value. The principal constraints are: temperature, electric field, magnetic induction, pressure and mechanical deformation. In examining these effects, we will supply the information needed to better understand the phenomena under consideration as well as their orders of magnitude.

According to solid-state quantum mechanics, charges, such as electrons in metals and metallic alloys, moving in a defect-free crystal lattice, do not collide with the lattice. If the periodicity of the crystal lattice is not absolutely perfect, however, there is a probability that the carriers will be diffused by the elements of the lattice. This activity, which Drude, for lack of a better word, referred to as the force of viscous friction (see Chap. 1), is responsible for the phenomenon of resistance. To calculate the value of this resistance we need to obtain the diffusion probability per collision and, consequently, the time between collisions 2τ and the mean free path λ.

The phenomena that introduce defects into the perfect periodicity of the crystal lattice can result from:

1. The movement of atoms around their mean position in the lattice, caused by an increase of thermal energy
2. The presence of different types of atoms in solid solution in the crystal lattice
3. The complete or partial absence of a lattice (amorphous or liquid state)
4. The presence of disordered zones at the grain boundaries
5. The presence of holes and interstitial defects in the lattice

2. MATTHIESSEN'S RULE[1]

One of the most important equations governing the mechanism of conduction in materials used to manufacture resistors is given by Matthiessen's rule. When the mechanisms of carrier diffusion are *independent of one another,* the total relaxation time τ can be expressed by the following equation:

$$\frac{1}{\tau} = \frac{1}{\tau_1} + \frac{1}{\tau_2} + \ldots \tag{1.1}$$

in which τ_1, τ_2, etc., represent the relaxation times associated with the various mechanisms of diffusion.

In solid solutions and dilute alloys, the diffusion of carriers by impurities, holes, interstitial defects, and grain boundaries is *independent of temperature.*

Equation 1.1 can then be written as follows:

$$\frac{1}{\langle \tau \rangle} = \frac{1}{\langle \tau_{T^\circ} \rangle} + \sum_{n=1}^{i} \frac{1}{\langle \tau_{\text{imp}} \rangle} \tag{1.2}$$

in which $\langle \tau \rangle$ is the total mean relaxation time, $\langle \tau_{T^\circ} \rangle$ the temperature-dependent relaxation time (electron-phonon interactions), and the second term the sum of all relaxation times, independent, or weakly dependent on temperature.

[1]Ziman, J. M., *Electrons and Phonons,* p. 285. Oxford University Press: 1962.

Mott, N. F. and Jones, H., *The Theory of the Properties of Metals and Alloys,* p. 286. Dover Publications, Inc.: 1958.

Equation 1.2 can then be written:

$$\rho_{\mathrm{TOT}} = \rho_{T^\circ} + \sum_{n=1}^{i} \rho_{\mathrm{imp}} \qquad (1.3)$$

Matthiessen's rule can only be applied through an approximation, by assuming that the diffusion resulting from collisions is isotropic. But it provides "good" orders of magnitude even when we assume that anisotropic collisions occur, and for a concentration of foreign atoms in the crystal sufficiently small to exist in solid solution.

Nevertheless, this equation is extremely useful for developing materials where $\tau_{T^\circ} \gg \sum \tau_{\mathrm{imp}}$, in which the change in conductivity as a function of temperature will be negligible.

We shall see further on, however, that this condition is sometimes difficult to bring about because of the activation energies of structures in which most of the diffusion occurs independently of temperature.

3. REVERSIBILITY ASSOCIATED WITH CHANGES IN TEMPERATURE

If we assume that each atom in the lattice vibrates with a maximum frequency ν_D about its equilibrium position, then, following Einstein's model for specific heat:

$$h\nu_D = k\Theta \qquad (1.4)$$

where Θ is the Debye[2] temperature, h is Planck's constant (6.6×10^{-34} Joule-sec), and k is Boltzmann's constant (1.3×10^{-23} Joule-deg^{-1}).

The binding forces among the atoms introduce a restoring force:

$$M \cdot \frac{d^2\varepsilon}{dt^2} + b\varepsilon = 0 \qquad (1.5)$$

[2]The Debye temperature is taken from the theory of the specific heat of solids. The "Debye approximation" assumes that the atomic lattice in a solid vibrates as a continuous medium, but that the vibration frequency cannot exceed a maximum value, chosen in such a way that the maximum number of modes is equal to the total number of degrees of freedom. We can thus define a "cutoff frequency" ν_D. To this frequency corresponds a temperature Θ, called the Debye temperature. Despite its simplicity compared with more complex theories for specific heat, the Debye approximation has the advantage of supplying a reliable order of magnitude for the vibrational energy of the atomic lattice in a solid, and, consequently, for various physical properties associated with it, such as IR absorption, electrical conductivity, solid-liquid transitions, etc. Readers interested in these questions should refer to J. M. Ziman, *op. cit.*

in which ε is the distance of movement of the atom from its equilibrium position and $-b\varepsilon$ is the restoring force.

The movement of carriers transmits energy to the atoms, which heats (electron-phonon interaction) the material (Joule effect) and increases its resistivity.

The study of interactive phenomena of this type enables us to demonstrate[3] that for high temperatures ($T > 200°K$) σ is proportional to:

$$\rho = 1/\sigma \approx \langle\varepsilon\rangle^2 \cong T/M\Theta^2, \quad T > \Theta \tag{1.6}$$

The resistance is proportional to the absolute temperature. With certain notable exceptions, for the transition metals especially, this equation is valid for the majority of the pure metals.

The logarithmic derivative as a function of the temperature of the previous expression, $\partial\rho/\rho\partial T$, called the temperature coefficient of resistance (TCR), has a value of approximately $4{,}000 \cdot 10^{-6}/°K$ *when the material (metal or alloy) is pure and free of crystal defects.*

The relaxation time varies between $8 \cdot 10^{-15}$ seconds for Li and $24 \cdot 10^{-15}$ seconds for Fe, which corresponds to a mean free path on the order of 200 to 400 nanometers.

The transition metals present exceptions to this rule. For alloys of Ag, Pd, Ni, and Pt, R decreases when T° increases within certain temperature intervals. This is due to the electron band's structure of these metals, in which s-band electrons can be transferred to empty levels in the d band, where they contribute to conductivity with increased mobility.[4]

We will show later on that this phenomenon is used to develop specially treated Ni-Cr or Cu-Ni alloys in which changes in σ with temperature are as small as possible. In practice, however, conducting materials possess impurities, crystal defects, and, when in a polycrystalline state, grain boundaries. Under these conditions the total resistivity of the material obeys Matthiessen's rule (Eq. 1.1).

In general, because of the electric charges associated with the great majority of crystal defects, impurities, and grain boundaries, the collision cross-sections between conducting electrons and such defects are much more effective than electron–phonon[5] interactions and, given the electro-

[3]Mott, N. F. and Jones, H., *The Theory of the Properties of Metals and Alloys,* p. 224. Dover Publications, Inc., New York: 1958.

[4]Ziman, *op. cit.,* chap. 9.

[5]Ziman, *op. cit.,* chap. 6, paragraph 3, p. 226.

static nature of the disruptive potential, the interaction is practically independent of, or only slightly dependent on, temperature.

We then conclude that:

$$\frac{d\rho_{\text{TOT}}}{dT^\circ} = \frac{d\rho_{T^\circ}}{dT^\circ} \tag{1.7}$$

On the other hand, the temperature coefficient given by the equation:

$$\text{TCR} = \frac{d\rho_{\text{TOT}}}{\rho_{\text{TOT}} \cdot dT^\circ} = \frac{d\rho_{T^\circ}}{\rho_{\text{TOT}} \cdot dT^\circ} \tag{1.8}$$

will be much smaller than that of a pure, perfect, and monocrystalline conducting material. This property is also exploited in the manufacture of resistors (basically thin-film devices) having very low temperature coefficients.

4. REVERSIBILITY ASSOCIATED WITH CHANGES IN ELECTRIC FIELD

In Chap. 1 we showed that Ohm's law is based on the assumption that the vector $\vec{J}$ is proportional to the field $\vec{E}$. In other words σ is independent of $\vec{E}$. In Ge and Si type semiconductors, measurements have showed that mobility, $\mu = q\tau/m$, is dependent on $\vec{E}$ by an equation of the following type:[6]

$$\mu = \mu_0(1 + \beta\vec{E}^2) \tag{1.9}$$

in which β is independent of $\vec{E}$.

This equation has been verified numerous times during measurements of semiconductors.[7] Its physical origin is based on the rate of increase of the kinetic energy of the carriers. In effect:

$$(dW/dt)_f = \text{force} \times \text{average velocity} = q\vec{E} \cdot \mu\vec{E} \tag{1.10}$$

In a steady-state condition this gives:

$$(dW/dt)_f + (dW/dt)_s = 0 \tag{1.11}$$

in which $(dW/dt)_s$ is the exchange of energy through electron–phonon diffusion.

[6]Conwell, E., "High Field Transport in Semiconductors," *Solid State Physics* Supplement, no. 9, chap. 2, p. 24 (1967), Academic Press.

[7]Gunn, J. B., *J. Physical Soc.*, Japan (1962): 1813.

This equation enables us to calculate the effective temperature T_e° of the carriers (this differs from the temperature of the crystal lattice T°). W. Schockley's calculation shows[8] that:

$$\frac{T_e^\circ}{T^\circ} = \frac{1}{2}\left[1 + (1 + Ct(\mu_0 E/\mu)^2)^{1/2}\right] \tag{1.12}$$

When the drift velocity qE is lower than the velocity of sound v_s in the crystal lattice ($v_s \approx 5 \cdot 10^3$ meter-sec^{-1}), we can write Eq. 1.12 in an approximate form as $T_e = T \cdot [1 + Ct(\mu_0 E/v_s)^2]$ and $T_e^\circ > T_c^\circ$. But this is a second-order effect and we can calculate that, for fields smaller than 10^4 V/M, $T_e^\circ = T^\circ$.

This effect, though less significant in materials generally used for resistors than in semiconductors, is nonetheless sufficiently large to be mentioned. It leads to the concept of voltage coefficient of resistivity. We should point out, however, that contrary to the temperature effect, which depends only on temperature, the voltage coefficient of resistivity is *highly dependent on the geometry of the resistor and especially on its length,* since in the great majority of cases, the electric field is parallel to the major axis of the resistor. Under these conditions it is easy to show that, based on $\Delta\mu/\mu = 2\beta\vec{E}d\vec{E}$ and Ohm's law $\vec{J} = \sigma \cdot \vec{E}$, the voltage coefficient of resistivity (VCR) is equal to:

$$\frac{\Delta R}{R \cdot \Delta V} = -2\beta \cdot \frac{V}{L^2} \tag{1.13}$$

in which V is the voltage applied across the terminals of a resistor of length L. The dimension of β is the inverse of the square of a field L^2V^{-2}, as predicted by Eq. 1.9. Its order of magnitude for materials that are frequently used in resistors is between 10^{-14} and 10^{-9} M^2V^{-2}. For a given resistor of width w and length L, we can see that, for a given voltage, the VCR is smaller when $L/w \gg 1$ (wound resistors, for example) than when $L/w \approx 1$ (thin-film resistors or cermets). In the majority of applications, this effect can be considered negligible when the materials involved are not semiconductors.

Measurement of the voltage coefficient is quite difficult because of the temperature coefficient, which generally masks its effects. Measurement consists in the application of a series of voltage pulses to the resistor. The cycle and width of the pulses are modified in such a way that, regardless of the value of the voltage, the power dissipated in the resistor is always the same.

[8]See for example, R. A. Smith, *Semiconductors,* chap. 5, p. 160, Cambridge University Press, 1959.

5. REVERSIBILITY ASSOCIATED WITH THE MAGNETIC FIELD

The Hall effect and magnetoresistance depend strongly on symmetry in the crystals of the materials under consideration. In this situation it is highly efficient to treat the problem from a phenomenological point of view and use the thermodynamics of irreversible processes. Using this approach, Ohm's law can be expressed[9] by the tensor equation:

$$E_i = \rho_{ij}(\vec{B})J_j + \varepsilon_{ijk}R_{Hk}(\vec{B}) \tag{1.14}$$

in which the resistivity ρ_{ij} and the Hall constant R_{Hk} are considered to be functions of the magnetic flux $\vec{B}$ applied to the material (ε_{ijk} is the rotation tensor such that $\varepsilon_{ijk} = 0$ if $i = j$, $i = k$, or $j = k$, and ± 1 in all other cases). The Onsager equations[10] show that ρ_{ij} is a symmetric tensor dependent only on even powers of $\vec{B}$, whereas R_{Hk} is an asymmetric tensor dependent solely on odd powers of $\vec{B}$.

ρ_{ij} and R_{Hk} can be developed through even- and odd-order MacLaurin series, respectively.[11]

In cubically symmetric crystal, whose electrical and mechanical properties are isotropic, the values of components of ρ_{ij} and R_{Hk} are independent of directions and are equal, respectively, to the mean material resistivity ρ_0 and Hall constant R_H. When the electric field is parallel to the x axis and magnetic induction is parallel to the z axis, Eq. 1.15 is reduced to the following equations:

$$E_x = J_x(\rho_0 + \xi_{xy}B_z^2 + \zeta_{xyy}B_0^4) + J_yB_z(R_H + \gamma_{xx}B_z^2 + k_{xxx}B_z^4) \tag{1.15}$$

$$E_y = J_y(\rho_0 + \xi_{xy}B_z^2 + \zeta_{xyy}B_z^4) - J_xB_z(R_H + \gamma_{xx}B_z^2 + k_{xxx}B_z^4) \tag{1.16}$$

$$E_z = J_z(\rho_0 + \xi_{xx}B_z + \zeta_{xxx}B_z^4) \tag{1.17}$$

If J_y^0 and J_z are zero, the first equation determines the transverse effect of magnetoresistance and the second the Hall effect. The third equation determines the longitudinal magnetoresistive constants. By neglecting second- and fourth-order effects in Eq. 1.16 and fourth-order effects in Eq. 1.17 we have:

$$E_y = -J_xB_zR_H \tag{1.18}$$

$$E_x = I_x(\rho_0 + \Delta\rho B_z^2 + \Delta\Delta B_z^4) \tag{1.19}$$

[9] Casimir, H. B. G., *Rev. Mod. Physics.* 17 (1945): 343.

[10] Onsager, L., *Physical Review.* 37 (1931): 405, ibid. 38 (1931): 2265.

[11] Mason, W. P., *Crystal Physics of Interaction Processes,* 247. Academic Press, 1966. Lovett, D. R., *Tensor Properties of Crystals,* p. 99, Adam Hilger, Bristol & Philadelphia.

There are numerous applications of the magnetoresistance and Hall effects. The measurement of the intensity of magnetic induction is one such application of the Hall effect. The magnetoresistance effect is primarily used in noncontact potentiometer sensors or more recently in heads for video cassette recorders and hard drives.

New developments in this field are now becoming known and finding industrial applications. Magnetic super-lattices, stacks of monocrystalline layers no more than several atoms thick in which magnetic and nonmagnetic materials alternate, exhibit enormous magnetoresistance. Their change in resistance when subject to the action of a magnetic field of a few tenths of a Tesla results in a change in resistance of between 50 and 100% at room temperature. The record magnetoresistances obtained with multiple layers of Co-Cu make these super-lattices the materials of choice for the construction of read heads for hard disks and magnetic tape and for high-definition television[12] VCRs. The phenomenon of magnetoresistance will be covered in greater detail in Chap. 12, which is devoted to its applications.

6. REVERSIBILITY ASSOCIATED WITH MECHANICAL FORCES AND DEFORMATIONS

We have seen that temperature is responsible for the amplitude of the vibrations of atoms in a crystal around their equilibrium position. This introduced Eq. 1.4, representing the change in resistance as a function of temperature. However, for a certain number of conductors, the variation of ρ as a function of T° is not exactly proportional. The transition metals and alloys do not obey this law. There are several reasons for this. One of them is due to the variation of Θ (Debye temperature) as a function of the change in the crystal's volume. According to Eq. 15 ($\rho/T \cong \Theta^{-2}$), we can write:

$$\frac{d\,ln(\rho/T)}{dT} = -2\frac{d\,ln(\Theta)}{dV} \cdot \frac{dV}{dT} \tag{1.20}$$

In 1912, Grüneisen[13] established a relation between the relative variation of Θ and the relative variation of the unit volume of a crystal:

$$\alpha_v = \frac{1}{V} \cdot \left(\frac{\partial V}{\partial T}\right) \quad \text{and} \quad G = -\frac{d\,ln(\Theta)}{d\,ln(V)} \tag{1.21}$$

[12] *Pour la Science.* (French edition of "Scientific American"): 61 (March 1991): 21.

[13] Grüneisen, E., *Ann. Phys.* 4, 39 (1912): 257.

in which α_v and G are respectively the volume coefficient of expansion and the Grüneisen constant. As a result:

$$\frac{d\log(\rho/T)}{dT} = 2\alpha G \tag{1.22}$$

By combining Eqs. 1.20 and 1.22, we can write an equation between the change in volume and resistivity:

$$\frac{\partial\, ln\, \rho}{\partial\, ln\, V} = \frac{\partial\, ln\, \rho}{\partial\, ln\, \theta} \cdot \frac{\partial\, ln\, \theta}{\partial\, ln\, V}, \text{ that is } \frac{d\rho}{\rho} = 2G \cdot \frac{dV}{V} \tag{1.23}$$

in which $2G$ is known as the Bridgman constant e.

Measurements have been made of conducting materials.[14] There are significant differences between the experimental results and theory for the transition metals. For the most part this is due to the structure of the s and d bands, which are superimposed on one another. This phenomenon of reversible drift due to the action of a pressure (force) is used, as we will show later on, in resistance strain gauges.

In practice, whenever an isotropic pressure is applied to a resistance $R = \rho \cdot L/l \cdot e$, maintained at constant temperature (L = length, l = width, and e the thickness of the resistor), we have:

$$\frac{dR}{R} = 2G \cdot \frac{dV}{V} + \frac{dL}{L} - \frac{dl}{l} - \frac{de}{e} = 2 \cdot \frac{dL}{L} + (\mathrm{e} - 1) \cdot \frac{dV}{V} \tag{1.24}$$

7. CONCLUSION

The effects of reversible phenomena on resistivity, and therefore on resistance, is one of the major factors affecting the performance of resistors. Depending on the intended use, a skilled practitioner will attempt to either minimize them, in the case of traditional electronic circuits, or maximize them. In the latter case, depending on the application, the resistors will be used primarily as temperature, field, or strain and stress sensors. We will later show that special materials, known as *varistors,* in which the phenomenon of the nonlinearity of Ohm's law as a function of electric field has been systematically maximized, are very common in circuit protection applications.

[14]Mott, N. F., *Proc. Phys. Soc.* 46 (1934): 680, England and Grüneisen, E. *Ann. Phys.* (5) 40 (1941): 543.

3

Irreversible Phenomena Associated with Ohm's Law

1. INTRODUCTION

For the most part irreversible phenomena are mechanisms that *permanently modify resistance*. There are two reasons for this:

1. a change in the resistivity of the material used
2. a change in resistance caused by phenomena external to the resistive material; for example, insulation loss, mechanical stress, etc.

In this chapter we will only describe those phenomena that permanently modify the resistivity of solid and thin-film metals and alloys. Cermets will be treated separately because of their great variety and the number of individual phenomena that give rise to conductivity (resistivity) in these materials.

Irreversible drifts in resistivity are caused by the effects of temperature and time on the structure of the resistive materials, and the effects of the corrosion caused during operation by applied electric fields.

2. TIME-TEMPERATURE EFFECTS

2.1 Kinetics and Diffusion Phenomena

In a solid solution composed of various types of atoms (A and B to simplify), the concentration c of atoms of a similar nature (A or B) can vary from one point to another in the solution. The spontaneous tendency of this system will thus tend toward uniformity. A flux of atoms will leave those portions of the solution that are rich in A or B for regions that are poor in A or B atoms. If we take a given direction x within the volume of the solution, we can define a *field of concentration* $\partial c/\partial x$ in the direction x. This field is related to the gradient concentration of particles created within the material by the variation in concentration of the said particles between adjacent volumes. The particles moving in the descending direction of the gradient and the flux J_x are proportional to the gradient such that:

$$J_x = -D \cdot \frac{\partial c}{\partial x} \tag{1.1}$$

D is called the diffusion coefficient, or diffusivity. It plays a role similar to that of electrical or thermal conductivity. The minus sign indicates that the diffusion tends to homogenize the mixture.

The flux is expressed as the number of particles crossing a unit of surface per unit time. It follows that the dimensions of D are L^2t^{-1}, generally expressed in $m^2 \cdot \mathrm{sec}^{-1}$. This law, known as Fick's law, is formally identical to Fourier's law for heat flow. It can be easily generalized to three dimensions:

$$\vec{J} = -D \cdot \overrightarrow{\mathrm{grad}}(c) \qquad (D \text{ is a second-rank tensor}) \tag{1.2}$$

In fact, the flow of particles in a given direction can arise from two sources: one is naturally the gradient of concentration, but the other is the *action of an outside force* or *transport force,* under whose influence the particles move with a given mean velocity $\langle v \rangle$, creating a flux $\langle v \rangle \cdot c$. Based on which:

$$J = -D \cdot \frac{\partial c}{\partial x} + \langle v \rangle \cdot c \tag{1.3}$$

The first term of the second member is a *diffusion* flux, whereas the second term is a *transport* flux.

In a transient state, when the flux at each point varies with time, Fick's law requires the addition of a continuity or conservation equation, which corresponds to a mass balance and connects the change in concentration per unit of time with the flux gradient:

$$\frac{\partial J}{\partial x} = -\frac{\partial c}{\partial t} \tag{1.4a}$$

By combining this with Fick's law, we obtain the equation for diffusion, which involves the use of second-order partial derivatives:

$$\frac{\partial c}{\partial t} = D \cdot \frac{\partial^2 c}{\partial x^2} - \langle v \rangle \cdot \frac{\partial c}{\partial x} \tag{1.4b}$$

This equation cannot be solved analytically as long as D and $\langle v \rangle$ depend on the concentration and, therefore, on x and t.

2.2 Nature of the Transport Force

The nature of the transport force can only be determined through the use of the thermodynamics of irreversible processes. We saw how in the presence of a charge transport an electric field gives rise to Ohm's law. There are other phenomena as well. Table 1 compares five examples of transport with the corresponding expression of force.

The first item in the table involves the presence of charged particles. Whereas for electrons or holes we speak of electrical conductivity, for atoms we speak of *electromigration,* or *electrotransport.* The charge q^* is that of the ion or point defect in ionic crystals. In a metal or alloy this charge can be very different from the charge of the ion.

Thermotransport, or *thermomigration,* is an important phenomenon when the materials are not in an isothermal state.

TABLE 1. Transport Forces

Type	Expression	Comments
Potential gradient	$q^*\vec{E}$	q^* = effective charge
Temperature gradient	$-\frac{Q \cdot dT^\circ}{T^\circ \cdot dx}$	Q = heat transmitted
Chemical potential gradient	$-kT^\circ \frac{\partial Ln(G)}{\partial x}$	G = chemical activation
Stress gradient	$\frac{dU}{dx}$	U = elastic potential
Centrifugal force	$m\omega^2 r$	

In the case of a *nonideal solution,* thc gradient of the activity coefficient introduces a transport force.

The effect of the *stress gradient* is noticeable whenever the elastic interaction energy of the particle is relatively large. This results in the migration of atoms in a sample that is subject to elastic deformation and the diffusion of interstitials, as well as the formation of a *Cottrel's atmosphere*[1], consisting of atoms surrounding a dislocation.

The effect of centrifugal force (for reference) can only be observed when very high velocities are involved.

2.3 Study and Determination of the Diffusion Coefficient

We have seen (refer to Chap. 2) that alloys can be described by phase diagrams correlated to changes in free enthalpy during these changes and phase transformations. However, velocity parameters required to achieve these transformations have not been considered or discussed.

This information is, however, basic to our understanding of the structures involved and consequently of the development of the properties of conductivity under constraints developed by their practical utilization. The same is true for other materials, such as thin films and cermets. In the great majority of cases, phase changes do not affect their thermodynamic equilibrium without the addition of energy to the system (segregation phenomena, reactions, incomplete oxidation). Yet, these materials often exist in metastable equilibrium (cermets and thin layers). To understand the phenomenon of irreversible drift, it is necessary therefore to introduce a theory, even a simplified one, of reaction kinetics and diffusion.

The reactions occurring during irreversible changes in conductivity imply, in the great majority of cases, that there is a movement of atoms or crystal defects in changing from one configuration to another, involving energy states with a clearly defined activation energy.

Let us assume that the free enthalpy of the system to which an ion (or atom) belongs can decrease by $\Delta G = G_2 - G_1 < 0$ if the ion moves from position 1 to position 2. The problem then becomes to calculate the change $G(x)$. In going from position 1 to position 2, this change generally involves a maximum that corresponds to the activated statc 1′. We therefore define $\Delta G^* = G_{1'} - G_1$ as the enthalpy of activation or energy of acti-

[1]Philibert, J., *Diffusion et Transport de Matière dans les Solides,* Chap. X, p. 430. Editions de Physique, 1985.

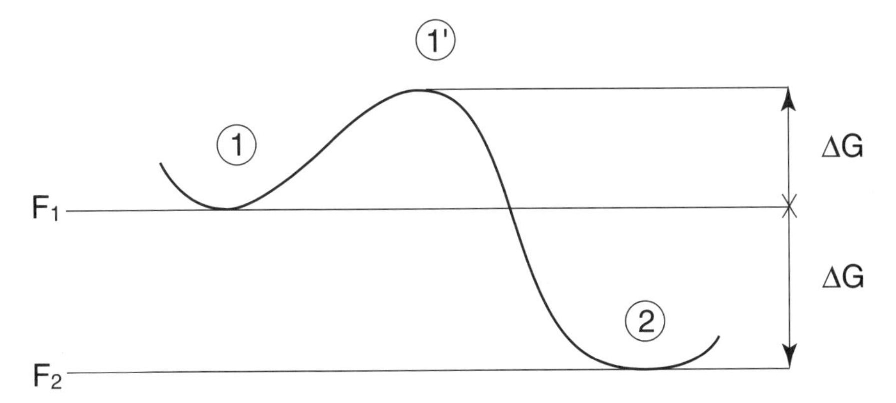

Fig. 1 The change $G(x)$ assumes the form of the curve shown above. There is a peak, or energy barrier, that must be overcome for the system to move from point 1 to point 2. It is obvious that the threshold velocity will depend on the "height" of this barrier, and thus of ΔG^*.

vation of the system. The change in $G(x)$ assumes the appearance of Fig. 1; it shows a peak, or energy barrier, that must be overcome to minimize the free enthalpy.

It is obvious that the probability of crossing the barrier will depend on the height ΔG^* of the barrier.

The "force" acting on the ion and which enables it to cross this barrier consists of the vibrations of frequency ν, which depend on the structure. We know that the mean energy of thermal agitation is proportional to kT, but its distribution is not uniform and follows Maxwell–Boltzmann statistics. At a given moment, some ions will have an energy greater than the threshold ΔG^*. The number of ions with this energy is proportional to exp $(-\Delta G^*/kT^\circ)$.

The number of jumps will obviously be proportional to:

$$\Gamma = \nu \cdot exp(-\Delta G^*/kT^\circ) \text{ in sec}^{-1} \tag{1.5}$$

Those processes that cross the activation barrier are said to be *thermally activated.* Their velocity depends on T° and increases with it. They govern all transformations in metallic alloys such as increase in grain size, tempering, case hardening, creep, etc.

This equation is known as the Arrhenius rate law.

A large number of failure mechanisms of electronic components are based on failure mechanisms that depend on Arrhenius-type phenomena: the diffusion of atoms to privileged sites in a crystal, the ability to overcome potential barriers, oxidation, etc. It would be useful, therefore, to be able to predict reliability based on specific tests. We can define a *temperature-time equivalence parameter.* Let us assume, to use a simple example, that the acceptable failure threshold can be determined by the number N of atoms that have diffused into the grain boundaries. The velocity of diffusion is $v = -dN/dt = \Gamma \cdot N$, in which N is the number of atoms that have undergone the transition $1 \Rightarrow 2$ at time t (assuming first-order kinetics). We can then write:

$$v = -\frac{dN}{dt} = \nu \cdot N \exp\left(-\frac{\Delta G^*}{kT^\circ}\right), \quad \text{therefore} \quad \frac{dN}{N} = \nu \cdot \exp\left(-\frac{\Delta G^*}{kT^\circ}\right) \tag{1.6}$$

If at temperature T_1 and time t, the failure threshold $S_\tau = 1 - 0.434\ Ln(N_1/N_0)$ has been reached (N_0 being the number of atoms that have not yet diffused at time $t = 0$), we can easily demonstrate that for a temperature T^0 and a time t, these two parameters are related by the equation:

$$P = Ln(S_\tau) = \left[\frac{1}{T^\circ} - \frac{kLn(t)}{\Delta G^*}\right] \tag{1.7}$$

This is the *time-temperature equivalence parameter.* If the phenomena satisfy the Arrhenius rate law and *if the activation energies do not vary with temperature* or *if an increase of temperature does not introduce any other activation energy term,* then knowing the transformation velocity at a given time t_1 and temperature T_1° enables us to predict the effects of a temperature T_2° at time t_2.

In reality, the processes are not always so simple. Very often there is competition between two or more reactions in parallel, as, for example, during the thermal repair of crystal defects and grain growth, or the diffusion of oxygen followed by oxidation. Figure 2 represents two reaction paths for the passage of an atom in position 0, 0, 0 to the vacancy position 1/2, 1/2, 1/2. It is obvious that with the slightest distortion of the lattice, path 1 will have a greater probability of occurring than path 2, but the two phenomena are equally present.

The orders of magnitude for ΔG^* and ν are respectively measured in tenths of an eV and from 10^{12} to $10^{14}\ s^{-1}$. If we select $\Delta G^* = 0.6\ eV$ and $\nu = 8 \cdot 10^{12}$ (representing a vacancy position for pure aluminum), then $\Gamma = 800\ s^{-1}$ at 300°K and $3.7 \times 10^9\ s^{-1}$ for 900°K. We see there is considerable acceleration of the diffusions with an increase in temperature.

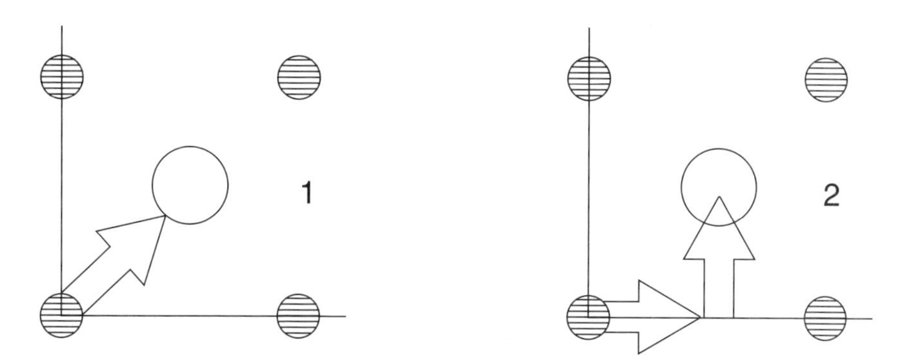

Fig. 2 Interstitial diffusion. The distortion of the crystalline structure is smaller in case 1 than in case 2. Consequently $\Delta G_1^* < \Delta G_2^*$ and case 1 will have a greater velocity.

In fact, the reactions that occur in resistive materials, regardless of type, depend on the process of germination, the movement and repair of crystal defects, and grain growth. It is very difficult to treat these very complex situations analytically and we must often resort to the use of empirical solutions involving smoothing or statistical regression of the experimental results. We could, for example, use a velocity equation of the following form:

$$\frac{dy}{dt} = \text{Const}^{\eta} \cdot t^{\eta-1} \cdot \exp\left(-\sum^{1} \Delta G_1/kT^{\circ}\right) \tag{1.8}$$

in which y can represent various properties of the material, including resistivity.

For example, take a simple cubic lattice which contains a radioactively tagged atom and where a represents the interplanar distance. Because this is not a very realistic example, since there is no known crystalline element with this lattice structure, we avoid introducing geometric complications in our model and arrive at a result that can be generalized to the lattices of metals and alloys (body-centered cubic, face-centered cubic, etc.).

After n jumps with a mean time $t = n/\Gamma$, therefore, the tagged atom has moved a vector $\vec{L}(t)$ from its point of origin. Because we are in an isotropic medium, the mean value $\langle \vec{L}(t) \rangle$ is zero. On the other hand, the *length of the displacement* is not zero on the average. The atom follows a random path whose distance is $n\lambda$ and whose standard deviation is $a \cdot (n)^{1/2}$. Therefore,

$$\langle L(t) \rangle^2 = na^2 = \Gamma a^2 t \tag{1.9}$$

Equation 1.9 represents the displacement of an atom based on its random movement. Consider the movement of this atom in a solid bar such that the x axis is oriented along the length of the bar's axis. The concentration will be uniform in each atomic plane, but it will vary from one plane to another. Although the atoms move by means of random jumps, the observable result is diffusion in one dimension; that is, between two atomic planes P_1 and P_2, which are contiguous with the abscissas x and $x + a$. An atom will jump from one plane to another with a frequency $\Gamma/6$ (one of the six elementary jumps possible). On surface P_1 the number of atoms is ca (the mean number of atoms per unit cell is ca^3 and there are $1/a^2$ unit cells per unit of surface). The number of atoms moving from P_1 to P_2 per second therefore is $c(x)a\Gamma/6$, and the number of atoms making the jump from P_2 to P_1 is $c\,(x + a)a\Gamma/6$. The two opposite currents do not neutralize one another if there is a difference in concentration between contiguous planes. The flux will therefore be:

$$J_x = a\Gamma/6[c(x) - c(x + a)] = -a^2\Gamma c'(x)/6$$

Comparing this result to Fick's law, we obtain an expression of the above coefficient such that:

$$D = \Gamma a^2/6 = \Gamma_s a^2 \tag{1.10}$$

The coefficient D is on the order of 10^{-25} $cm^2 \cdot s^{-1}$ at room temperature, but increases very rapidly with temperature (10^{-12} at 500–600°K).

In the presence of a transport force $\vec{F}$, Γ will obviously have a different value depending on the application direction of the force. In other words, $\Gamma_{1\Rightarrow 2} \cong \Gamma_{2\Rightarrow 1.}$

We can write a limited expansion:

$$\vec{F}_{1\Rightarrow 2} = \vec{F}_s(1 + \varepsilon) \tag{1.11}$$

$$\vec{F}_{2\Rightarrow 1} = \vec{F}_s(1 - \varepsilon) \tag{1.12}$$

based on which

$$\langle \nu \rangle = a \cdot (\vec{F}_{1\Rightarrow 2} - \vec{F}_{2\Rightarrow 1}) = 2\varepsilon a\vec{F}_s \tag{1.13}$$

Expressing the net thermal energy kT per jump in the direction of the applied force corresponding to the work exerted by that force over distance λ, we get:

$$kT°(\vec{F}_{1\Rightarrow 2} - \vec{F}_{2\Rightarrow 1})/\vec{F}_s = \vec{F} \cdot a \tag{1.14}$$

$$\varepsilon = (1/2)\vec{F} \cdot a/kT \tag{1.15}$$

and

$$\langle \nu \rangle = \vec{F}D/kT \text{ (Nernst–Einstein equation[2])} \qquad (1.16)$$

2.4 Diffusion Mechanisms

In our discussion so far we have only examined the case of an interstitial diffusion as shown in Fig. 2. This diffusion occurs much more easily when the diffusing atom is small in comparison to the atoms in the lattice. Carbon, a small atom, diffuses through steel in this manner. In fact, C, O, N, B, and H undergo interstitial diffusion in the majority of crystals.

The second mechanism is vacancy diffusion. When zinc diffuses through brass, for example, the zinc atom, similar to copper in size, is unable to find an interstitial site. The zinc must wait for the appearance of a *vacancy,* that is, an unoccupied crystalline site, in its neighborhood before it can diffuse. This is the most frequent mechanism of diffusion in crystals (Fig. 3).

Interstitial and vacancy diffusion in crystals can sometimes be *short-circuited* by diffusion along grain boundaries or dislocations. Grain boundaries behave like flat channels, two interatomic distances in width, with a local diffusion coefficient that can be, all other factors being equal (such as temperature), 10^6–10^7 times greater than the interstitials and vacancies, whose D values are of the order of 10^{-18} cm$^2 \cdot$ sec^{-1} (Figs. 4 and 5). The center of a dislocation can also behave like a conducting "wire" for diffusing atoms, with a cross-section of the order of $(2a)^2$, where a is the interatomic distance between the grains or dislocations. If the diameter of the grains is small and there are many dislocations, their contribution to the total diffusion flux can be significant.[3]

The movement of atoms around dislocations and grain boundaries in crystals is used to measure creep, an important phenomenon in strain gauges.

These competing mechanisms lead to deviations from the Arrhenius rate law and it is often difficult to determine their relative effects in a given crystal. For this reason there are very few situations in which Eq. 1.5, using a single activation energy, is sufficient to explain and predict irreversible changes in resistivity.

[2]Philibert, J., *op. cit.,* p. 31.

[3]Geiger, G. H. and Poirier, D. R., *Transport Phenomena in Metallurgy.* Chapter 13. Addison-Wesley, 1973.

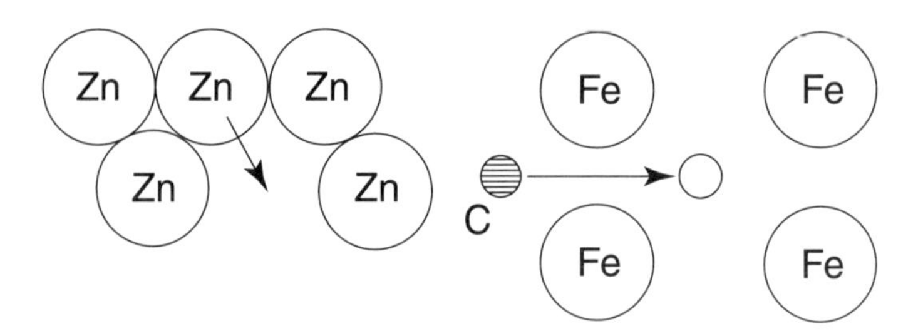

Fig. 3 Vacancy and interstitial diffusion.

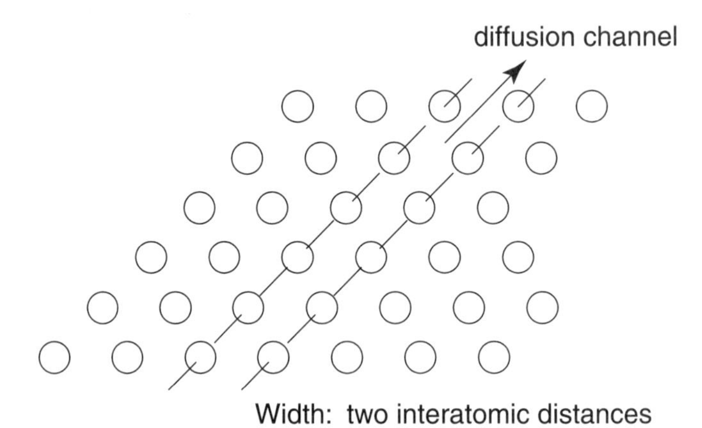

Fig. 4 Diffusion at grain boundaries.

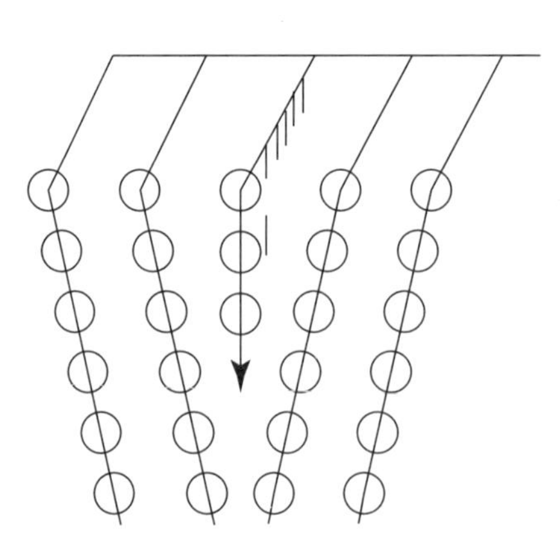

Fig. 5 Diffusion of dislocations.

Aside from irreversible phenomena in resistive materials, there are many phenomena governed by diffusion. The majority of them are used in manufacturing processes. These include:

Annealing homogenization
Phase transformation
Precipitation
Creep plastic deformation
Oxidation

2.5 Permanent Changes in Resistivity

In terms of conductivity, if the crystals under investigation are perfect and monocrystalline, there could be no irreversible phenomena associated with Ohm's law since the lattice atoms would be indistinguishable.

This is obviously not the case, however, and conductivity will vary as a function of time and temperature through Matthiessen's law (Chap. 2, Eq. 1.2).

Diffusion caused by vibrations of the crystal lattice at temperatures below 800–1000°K is responsible for temperature-dependent conductivity; it is isotropic by definition. On the other hand, diffusion or collisions with crystal defects such as interstitials, vacancies, and dislocations are not isotropic.

If σ_t is the total collision cross-section of the electron-defect interaction, the mean free path λ is given by $1/\lambda = N_{\text{def}}\sigma_t$ in which N_{def} is the number of defects per unit volume. The relaxation time τ_{imp} is given by:

$$\frac{1}{\tau_{\text{imp}}} = N_{\text{def}} v_F \sigma_t \tag{1.17}$$

in which v_F is, as was shown in Chap. 1, the electron velocity at the Fermi level.

From Matthiessen's rule, each defect will contribute to the resistivity (conductivity) according to the following formula:

$$\rho_{\text{TOT}} - \frac{m}{Nq^2} \cdot \frac{1}{\tau_{\text{TOT}}} = \frac{m}{Nq^2} \cdot \left[\frac{1}{\tau_{T^\circ}} + \sum^{1} \frac{1}{\tau_{\text{imp}}} \right] \tag{1.18}$$

That is:

$$\rho_{\text{TOT}} = \frac{m}{Nq^2} \cdot \left[\frac{1}{\tau_{T^\circ}} + \sum^{1} N_{\text{def}} V_F \sigma_1 \right] \tag{1.19}$$

to which are added effects "external" $\sum N_{ext}$ to the material, such as oxidation, nitriding, and corrosion phenomena.

We see that the resistivity will depend on the sum $\sum N_{def}$ of the defects, the number of atoms in solution, and the number of dislocations per unit volume in the material. This parameter being a function of temperature and time as shown by the equations developed previously, the resistivity will undergo an irreversible drift.

These effects are far from negligible. In fact, the vacancies and interstitials have a $\Delta\rho_{vac,int}$ on the order of $4 \cdot 10^{-19} \cdot N_{def}$ in a metal such as Cu or Ni.

It is generally acknowledged that the number of such vacancies and interstitials in a metal or alloy that has been work-hardened and then annealed is on the order of 10^{-13} to 10^{-15} cm^2, which results in a very significant contribution to the total resistivity.

To this must be added the influence of grain boundaries and dislocations, for which we estimate that $\Delta\rho$ is roughly $6\text{–}7 \cdot 10^{-17} \cdot N_{defects,boundaries}$ ohm · cm.

The influence of external parameters is also very significant, since an atom of oxygen removes one or more conduction electrons from the material. This will have an effect on $\sum N_{def}$ through distortion of the crystal and the introduction of interstitials, but also on the number of carriers per unit volume N.

The variation in time of $\sum N_{def}$ will follow Fick's laws.

Crystal defects will be eliminated through diffusion along grain boundaries and along the surfaces and boundaries that form "wells" (vacancy defects will also be created at temperature equilibrium). On the other hand, the same surfaces will become "sources" for "external" effects.

Two examples will help illustrate the above concept. We will assume a resistor composed of a thin-film material of thickness 1. The surfaces of this film form the defect-suppression wells.

Figure 6 represents the dynamics of eliminating vacancy saturation at the surface of a resistive film of thickness d. At time t_0 the vacancy density N_d within a given thickness is represented by curve 0, which changes with time (and temperature) in the direction of curves 1, 2, and 3. Curves 2 and 3 represent the density N_d at times t_1 and t_2, and temperature T_1°. This "healing," if it is the only factor involved, will introduce a negative drift in the resistance.

The solution of the diffusion equation in a transient state is obtained in this case by separating the variables. The solution assumes the form $N_d(x, t) = X(x) \cdot T(t)$ and Fick's equation becomes:

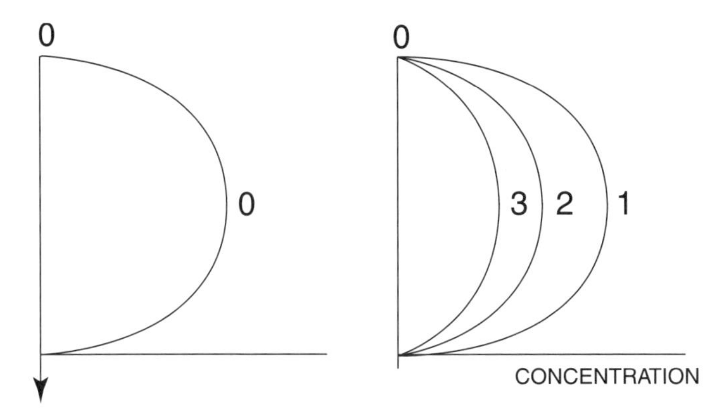

Fig. 6 Change in time of elimination of vacancy saturation.

$$\frac{1}{T}\frac{dT}{dt} = -\zeta^2 D \qquad \text{and} \qquad \frac{1}{X}\cdot\frac{d^2X}{dx^2} = -\zeta^2 \tag{1.20}$$

The solutions to these equations can be written as:

$$T = \exp(-\zeta^2 Dt) \qquad \text{and} \qquad X = A\sin(\zeta x) + B\cos(\zeta x) \tag{1.21}$$

The initial and limit conditions determine a series of values ζ_n, A_n, and B_n. In Fig. 6 we have $N_d(x, 0) = N_o$ and $N_d(x, t) = 0$ for $x = 0$, $x = d$, and $t > 0$. Under these conditions:

$$\frac{N_d(x,t)}{N_o} = \frac{4}{\pi}\cdot\sum_{n=1}^{\infty}\frac{1}{(2n+1)}\exp\left[-\frac{(2n+1)^2\pi^2}{d^2}\cdot Dt\right]\sin\left[\frac{(2n+1)\pi x}{d}\right] \tag{1.22}$$

The terms where $n \geq 0$ decay very quickly. We can approach the solution to times that are not too short, through:

$$\frac{N_d(x,t)}{N_o} = \frac{4}{\pi}\cdot\exp\left[-\frac{\pi^2}{d^2}\cdot Dt\right]\sin\left[\frac{\pi x}{d}\right] \tag{1.23}$$

in which D depends on the temperature as shown by the above equations. The number of vacancies still present at time t per unit of surface is $N_L(t) = N_{LO}\int_0^1 N_d(x,t)$ in which $N_{LO} = N_o d$. That is, by neglecting the transients:

$$N_L(t) \cong N_{LO}\exp\left[-\frac{\pi^2}{d^2}Dt\right] \tag{1.24}$$

A similar approximation leads to the same laws for cylindrical shapes.[4]

$$\frac{N_L(t)}{N_{LO}(O)} = \exp\left[-\frac{5.784Dt}{R^2}\right] \tag{1.25}$$

This equation can be applied to wound resistors in which the resistive element is a wire.

Dislocations and segregations follow the so-called Cottrell–Bilby laws.[5] That is,

$$N(t) \cong Ct \cdot D^{2/(n+2)} t^{2/(n+2)} \tag{1.26}$$

When the size effect is dominant (for example, thin-film media), $n = 1$ and the system kinetics vary with $t^{2/3}$.

2.6 Oxidation of a Metal or Alloy

Oxidation involves the diffusion of an element from a gaseous atmosphere together with a chemical reaction involving the formation of a compound. The same considerations apply when the diffusing element belongs to column 6 (oxygen, sulfur) or 7 (halogen), or to other elements supplied by a gas whose molecules contain that element (methane, ammonia, borane, silane, etc.)

Figure 7 represents the growth of an oxide layer on a metal or alloy *M*, the diffusion of oxygen occurring through the presence of vacancy defects. The oxide layer displays a composition gradient; that is, a gradient with deviation *d* from stoichiometry.

The chemical diffusion coefficient comes into play during chemical reactivity with a gas (oxidation, sulfidization, etc.) and governs the growth of the layer of the reaction product (oxide, sulfide, nitride, etc.) of a metal or alloy in a reactive atmosphere.

The oxidation mechanisms are expressed by four laws of growth. The depth of oxidation follows a parabolic, linear, logarithmic, or cubic law.

Metals that appear in the materials used to manufacture resistors (Cu, Fe, Co, Ni, etc.) generally obey a parabolic growth law. The rate of growth of the oxide layer is inversely proportional to the thickness of the previously created oxide layer and proportional to a constant K_0, which is characteristic of each metal:

$$\frac{dz}{dt} = \frac{K_0}{z} \tag{1.27}$$

[4]Philibert, J., *op. cit.*, p. 423.

[5]Cottrell, A. H. and Bilby, B. A., *Proc. Phys. Soc.* London. A 62, (1949): 49–62.

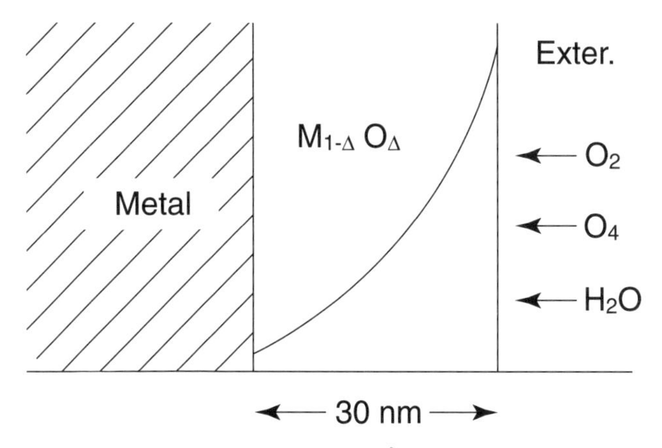

Fig. 7 Oxidation of a metal.

from which we derive:

$$z = (2K_0 t)^{1/2} \tag{1.28}$$

in which z is the thickness of the oxide layer at time t and K_p is an *oxidation constant* (in $L^2 \cdot s^{-1}$). As a first approximation, the constant K_0 is proportional to the difference in composition of the oxide between the outside surface and the metal/oxide interface, and to the coefficient of diffusion of the oxide in the metal.[6]

We saw in Chap. 2 that metallic resistive elements in resistors were often made of binary alloys of Ni–Cr or Ni–Cu. Because the stoichiometry is one of the important parameters governing performance, it will be worthwhile studying the conditions under which this stoichiometry varies during oxidation. In an alloy of type *A–B*, *A* can oxidize easier than *B* or vice versa. In the case of Ni–Cr alloys, chromium oxidizes more rapidly than Ni and the material will undergo a phase change. We must, therefore, look for solutions to Fick's equations that satisfy the specific constraints of the change of phase. In the case of Ni–Cr, since Cr oxidizes more easily, the oxidized layer grows rich in Ni. In this case we can demonstrate[7] that the concentration of Ni is expressed by the following equation:

[6]Jastrebski, Z. D., *The Nature and Properties of Engineering Materials*. p. 492. J. Wiley & Sons, 1977.

Philibert, J., op. cit., p. 207.

[7]Philibert, J., *op. cit.* pp. 20–24, 387–391.

$$c(z,t) = B1 + A1 \cdot \mathrm{erf}(z/2\sqrt{Dt}) \tag{1.29}$$[8]

in which D is the coefficient of heterodiffusion of Ni in Cr.

The oxide–metal interface obeys the following law:

$$\xi = 2C\sqrt{Dt} \tag{1.30}$$

$A1$, $B1$, and C depend on the limit conditions at the surface for $t = 0$ and at the interface ξ for $t > 0$ (Fig. 8).

To calculate the effect of these disturbances on the resistance drift as a function of time-temperature, time-oxidation, and other parameters, the values in the equations (of c, etc.) are introduced in the equation (of ρ) and the change calculated for a given resistor geometry by applying the integral Ohm's law as shown in Chap. 1.

However, given the nature of the phenomena described above, we see that it is very difficult to predict irreversible changes in resistance in conducting materials. These changes are a function of the various and competing mechanisms (healing, diffusion, oxidation, etc.) that occur. The relative size of these effects will depend strongly on the history of the material under consideration as well as its environment. Similarly, an analysis of fundamental causes based on experimental results should make use of several measurements and correlations between the measured drift and experimental conditions. This is further complicated by the fact that the drift is also the result of interface variations and contact resistance, as we will show in the chapter on the "environment."

Nevertheless, we can predict a number of different types of behavior: Resistors made from Ni–Cr or Ni–Cu alloys, whose geometry is such that surface effects can be neglected compared to volume, will have negative irreversible drift compared to the value of the resistance at time $t = 0$. This is basically due to the arrangements of the crystal and defect elimination (at least if the applied temperature is not high enough to result in phase changes).

These materials will thus behave in accordance with the Arrhenius and Fick laws. This can be corroborated from the positive drift in TCR, which will vary in relation to Matthiessen's rule.

Similarly, resistors manufactured from thick-film cermets, heterogeneous materials "packed" with defects, as a first approximation will conform to the laws of Arrhenius and Fick, and their activation energies can be determined from their behavior in response to changes in temperature and time.

[8]This function is called an "error function." It is expressed by the equation $\int_0^t \varphi(x)dx = 1/2\,\mathrm{erf}(t/\sqrt{2})$, in which $\phi(x)$ is the normal Gaussian function. It can be found in standard tables of mathematical formulas.

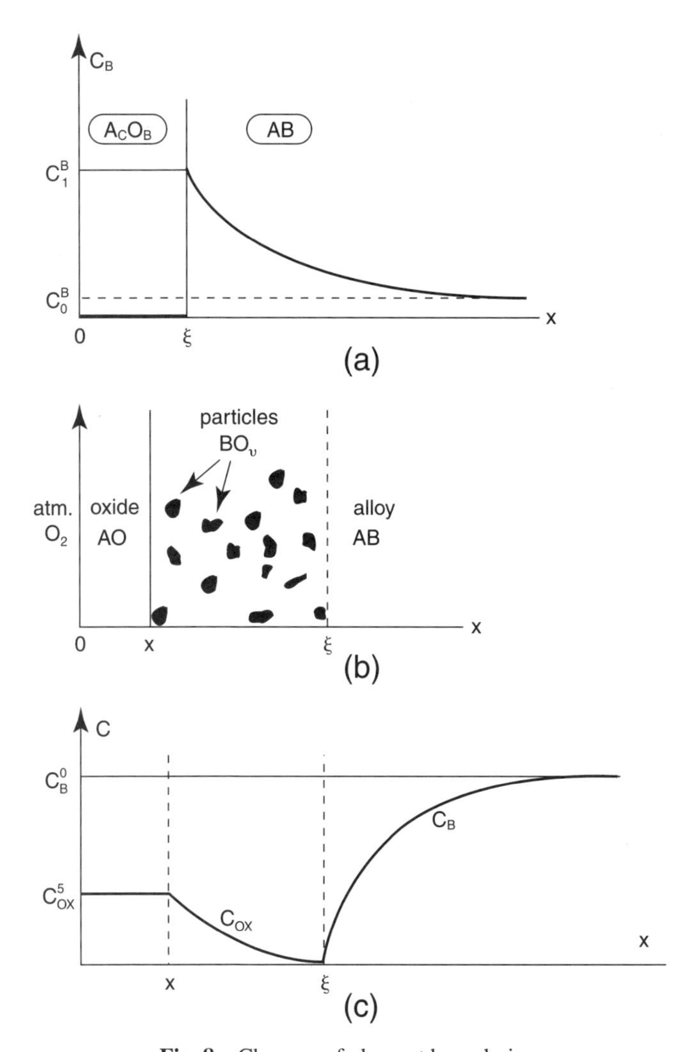

Fig. 8 Changes of phase at boundaries.

The behavior of thin-film materials (thickness ≤10,000 nm), however, will be governed primarily by defect healing, giving rise to negative drift. This drift quickly becomes positive (after several hours at 425°K) however, since it will then be governed by oxidation phenomena. If the changes introduced from external sources such as contact resistance and mechanical effects are sufficiently weak, the TCR of the resistor or material will drift positively and then negatively.

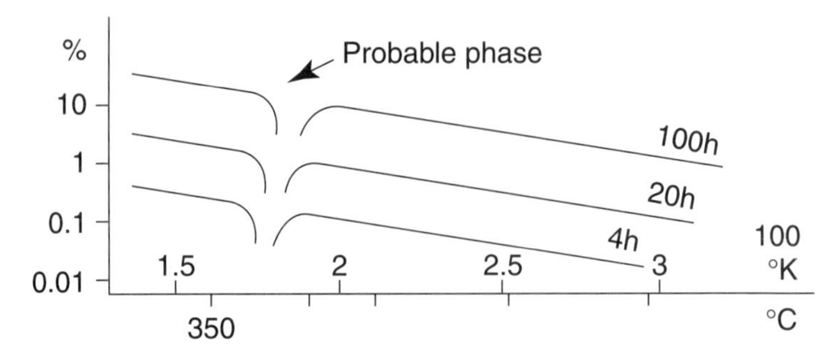

Fig. 9 Variation in resistivity of an Ag-Pd-Pd0 cermet as a function of temperature and time. Illustration of the Arrhenius rate law.

Figure 9 represents the shift in resistivity of an Ag-Pd-PdO cermet layer as a function of temperature and time. The shape of the curve shows, to a good approximation, that the aging of thick-layer cermets follows an Arrhenius rate law. The curve shows a singularity around 550°K, however, probably related to a phase transformation in the glass or alloy.

4

Resistive Materials

1. INTRODUCTION

Conductive materials used to manufacture resistors can be classified into three main categories:

(a) Metals and solid alloys in thermodynamic equilibrium. Their electrical properties will depend on the intrinsic crystalline structure of these materials and the metallurgical methods employed in their production.

(b) Thin-film materials obtained through evaporation, sputter deposition, electrolysis, and chemical vapor deposition of solid materials of the same nature as those above. Their electrical properties will, for the most part, be dependent on the impurities introduced during their formation, conditions of thermodynamic disequilibrium compared to their bulk structure, and their thickness (from a few nanometers to several micrometers).

(c) Very heterogeneous structures in the form of disordered mixtures of conventional conducting materials, dielectrics, precious metals (Ag, Au, Pt, etc.), or oxides. Their electrical properties will no longer depend on the crystal structure of their components but primarily on the random and topological nature of the mixtures.

2. METALS AND SOLID ALLOYS

The alloys commonly used in wirewound power, wirewound precision, and bonded rolled foil resistors are of the Cu-Ni type (constantan), Cu-Mn-Ni (manganin), and Ni-Cr.

2.1 Copper-Nickel and Copper-Manganese-Nickel Alloys

Cu-Ni alloys form a single substitutional solid solution in a face-centered cubic lattice for all compositions (Fig. 1).

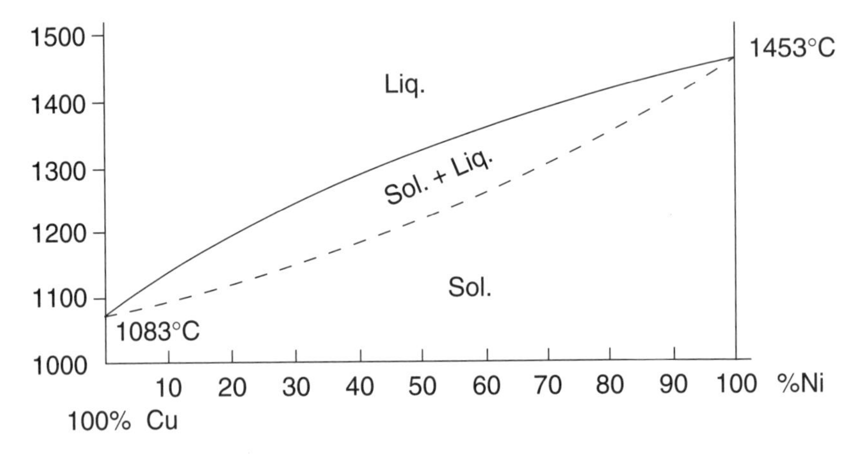

Fig. 1 Cu-Ni phase diagram.

A copper atom contains one electron more ($Z = 29$) than nickel ($Z = 28$).[1] Consequently, if a nickel atom is replaced by a copper atom, we can assume—and this turns out to be the case in practice—that the lattice will be unchanged and that a disordered solid solution will be obtained. Figure 1 shows the variation in resistivity as a function of composition. However, at a concentration of around 50% (atomic composition) nickel, X-ray diffraction as well as magnetic anomaly studies suggest the existence of a Ni-Cu super-lattice that exhibits nearest-neighbor ordering. The alloys used to manufacture resistors are selected from mixtures similar

[1] Z represents the atomic number of the material, which is equal to the number of electrons orbiting the nucleus of the neutral atom. The atomic weight represents the sum of all the particles contained in the neutral atom.

to this. The electrical characteristics of the alloy with around 50% Cu are the following:[2]

Resistivity: 60–65 $\mu\Omega \cdot$ cm TCR: $\pm$150 ppm/°C

This alloy, which has long been used for wirewound power resistors, has now been replaced in many applications by Ni-Cr alloys.

It is also used to create electro-welded ohmic contacts, using very thin (a few tens of μm) laminated ribbon.

Ternary alloys of Cu-Mn-Ni generally consist of Cu (84% atomic composition) —Mn (12% atomic composition) —Ni (4% atomic composition). In this form, their electrical characteristics are as follows:[3]

Resistivity: 50 $\mu\Omega \cdot$ cm TCR: $\pm$ 150 ppm/°C

Used for a long time to manufacture wirewound resistors, constantan and manganin have been supplanted by Ni-Cr alloys. Because of the weakness of the Seebeck effect in Ni-Cr alloys compared to copper, they have been used as terminal wires in resistors. Unfortunately, this characteristic is accompanied by poor thermal conductivity (40 $W \cdot m^{-1} \cdot K^{-1}$ compared to 400 for copper). As a result of the progress that has been made in optimizing heat transfer, and the weak temperature gradient observed in recently developed resistive components, they have been replaced by copper wire or ribbon.

2.2 Nickel-Chrome Alloys

At room temperature the phase diagram for nickel-chrome alloys (Fig. 2) consists of only two solid solutions, referred to as α and γ by analogy with ferrite and austenite in the iron-carbon diagram. The first, similar to

[2]The temperature coefficient of resistivity, TCR, of a resistor is the relative change in its value as a function of temperature. It is determined by the equation TCR $= d\rho/\rho dT$ for the resistivity of the material and TCR $= dR/RdT$ for the resistive component. Chapter 2 describes in detail the mechanisms of conduction on which the TCR is based.

[3]Furukawa, Reilly, and Saha., *Rev. Sci. Instr.* 35 (1964): 113.

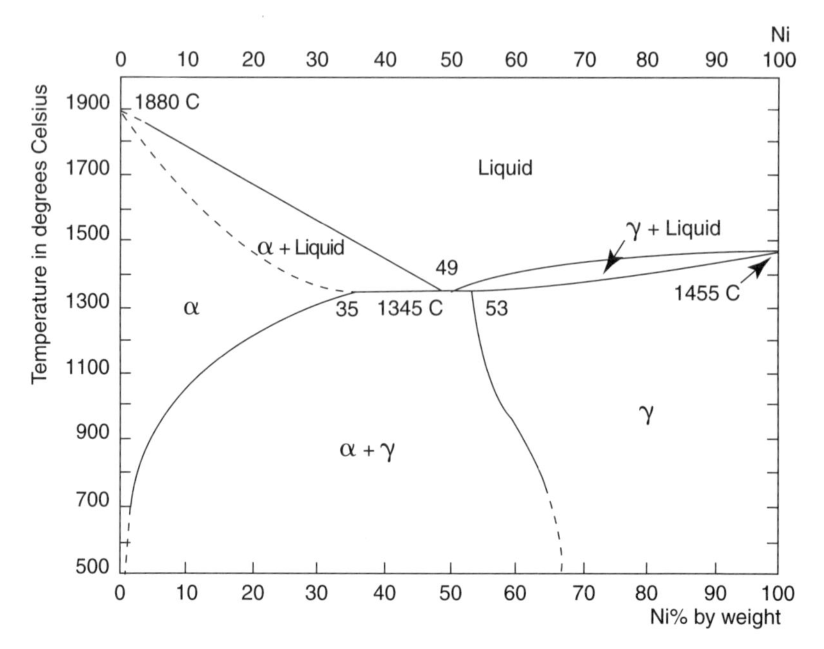

Fig. 2 Nickel-chromium diagram (Hansen). Dashed line: incompletely known curves.

chrome, has a cubic-centered crystal lattice, whereas the second, similar to nickel, has a face-centered cubic lattice.

Because of the slope of the boundary between the single-phase γ domain and the two-phase domain, alloys with approximately 60% Ni, when cooled slowly, precipitate the α phase in the form of parallel sheets in the γ solution. Unless they are quenched, a similar precipitation of the γ solution in the α solution can be observed in chromium-rich alloys, the boundary between the single-phase α domain and the two-phase domain being particularly far from the vertical.

The most commonly used nickel-chrome alloys are found in the single-phase γ domain.

The most common alloy is made of 80% (atomic composition) nickel and 20% (atomic composition) chromium. This alloy maintains its mechanical strength up to high temperatures and is very resistant to heat oxidation.

The alloy is used as a base material for wirewound power resistors and foil resistors, its high resistivity and sufficiently weak TCR (less than

±150 ppm/°C) making it a very suitable material. Nevertheless, certain operating conditions require that complementary metals (Cu, Mn, Al, etc.) be added.

In general, the alloy will deteriorate through local flaking of the protective oxide film. It is important to provide this film with good adhesive properties, which is accomplished through the addition of very small amounts of calcium, cerium, or zirconium.

The use of this alloy in high-precision[4] resistors requires that, in order to reduce the TCR by one of the mechanisms explained earlier, part of the chromium be replaced by iron or copper, and part of the nickel by aluminum, which also provides increased resistance to creep.

The result is a quaternary alloy containing 77% (atomic composition) nickel, 15% (atomic composition) chromium, 5% (atomic composition) iron or copper, and 3% (atomic composition) aluminum. Depending on the manufacturer and the variance in Fe or Cu content from the stated values, the alloy is sold under the names Karma,[5] Moleculoy, etc.

Following appropriate thermal treatments to remove any structural defects introduced during manufacture (rolling, etc.), the electrical properties of this alloy are as follows:

Resistivity: 130 $\mu\Omega \cdot$ cm
TCR: 0 ppm/°C at 25°C ± 3 ppm/°C between −55° and 125°C
and in some cases as low as 1 ppm/°C (foil resistors)[6]

2.3 Conduction Mechanism in Magnetic Alloys

We have seen that the Cu-Ni and Ni-Cr alloys with traces of Fe, Cu, or Al are characterized by high resistivity and very low temperature coefficients, which contradict the electronic theory of conduction in metals

[4]Manufacturers and users traditionally characterize high-precision resistors as having tolerances on the order of 0.05% of their rated value, a TCR less than +5 ppm/°C, irreversible drift below +500 ppm at 125°C, and 10,000 hours of operation. This usage has been well established by the profession and is reflected in the official standards.

[5]Furukawa, G. T. *et al., Rev. Sci. Instr.* vol. 35 (1964): 113, American Physical Society.

[6]A number of measurements by alloy producers and resistor manufacturers have shown that the representative TCR curve for this family of alloys as a function of temperature has the form: $\Delta R/R_0 = A(T - T_0) + B(T - T_0)^2 + C(T - T_0)^3$ in which $T_0 = 25$°C, R_0 is the resistance at 25°C, and A, B, and C are constants that are characteristic of the material, such as

and alloys discussed in Chap. 1. It would, therefore, be worthwhile to examine the specific mechanisms of conduction in the transition metals and their alloys. The specific properties of the transition metals are associated with the existence of $3d$ shells, which are gradually filled before the $4s$ shell when atomic number increases. This phenomenon is corroborated by X-ray scattering and by optical measurements (change in reflectivity), as well as by measurements of specific heat, magnetic susceptibility, and finally, resistivity.

The high resistivity of these metals and alloys can be interpreted as resulting from the high density of d states at the Fermi level,[7] unlike ordinary[8] metals. This density explains the ferromagnetic behavior of these metals and alloys. More specifically, the model can be summarized as follows:

1. Conduction is basically provided by s-band electrons. In effect, if we assume the existence of two independent bands, the total conductivity σ is the sum of the conductivities associated with the d and s bands:

$$\sigma = \sigma_d + \sigma_s \text{ in which } \sigma_i = \frac{\Re_i q^2 \tau_i}{m_i} \tag{1.1}$$

in which $\Re_i$ is the number of electrons s or holes d per atom, τ_i the relaxation time at the Fermi energy, and m_i the effective mass of the electrons in the s and d bands, both of which are assumed to be parabolic. Because the d bands are narrow and nearly filled, it follows that $n_d(E_F) >> n_s(E_F)$, $m_d >> m_s$ and $\sigma_d << \sigma_s$.

2. The high resistivity and weak dependence of the temperature coefficient TCR on temperature are governed by $s \Longleftrightarrow d$ transitions. Electron-phonon or electron-impurity diffusions caused by the perturbing potential in the lattice cause transitions between a conducting s state and nonconducting d states, with a probability that is dependent on the perturbing potential, n_d, and the spin state of the electron.[9]

$A \cong 1 \pm 1.10^{-6}$, $B \cong 4.10^{-8} \pm 1.10^{-9}$ and $C \cong 10^{-10}$. The respective values of these constants vary within specific intervals depending on the heat treatment applied and the impurities present in the alloy.

[7]Mott, N. F. and Jones, H., *The Theory of the Properties of Metals and Alloys. op. cit.*

[8]Janot, C. *et al., Propriétés électroniques des métaux et alliages.* Ecole d'Eté de Royan. Masson & Cie, France, 1971.

[9]Fert, A. and Campbell, I. A., *J. de Phys. F: Met. Phys.* 6 (1976): 849.

In other words, the *d* band forms a trap into which the *s* electrons are diffused and where they no longer contribute to conduction. This is confirmed in detail from conductivity studies on alloys of transition metals with monovalent metals. The Pd-Ag alloy is typical in this respect. For a given composition, the conduction electrons no longer exist in sufficient numbers to fill the *d* band in the palladium, and the resistance increases sharply. This is also the case for the Cu-Ni alloy when the Ni content is 40% atomic composition or greater (Fig. 3).

Moreover, because of the spin of the *s*- and *d*-band electrons, scattering is considered inelastic through application of the Pauli exclusion principle. As a result the conduction electrons are much more closely bound to the form of the free-energy states near the Fermi level than in the classical model, and some electron-phonon scattering is forbidden to them. This phenomenon is the source of the TCR anomalies in transition metals and their alloys, and in particular of the low value of the TCR around 300°K.

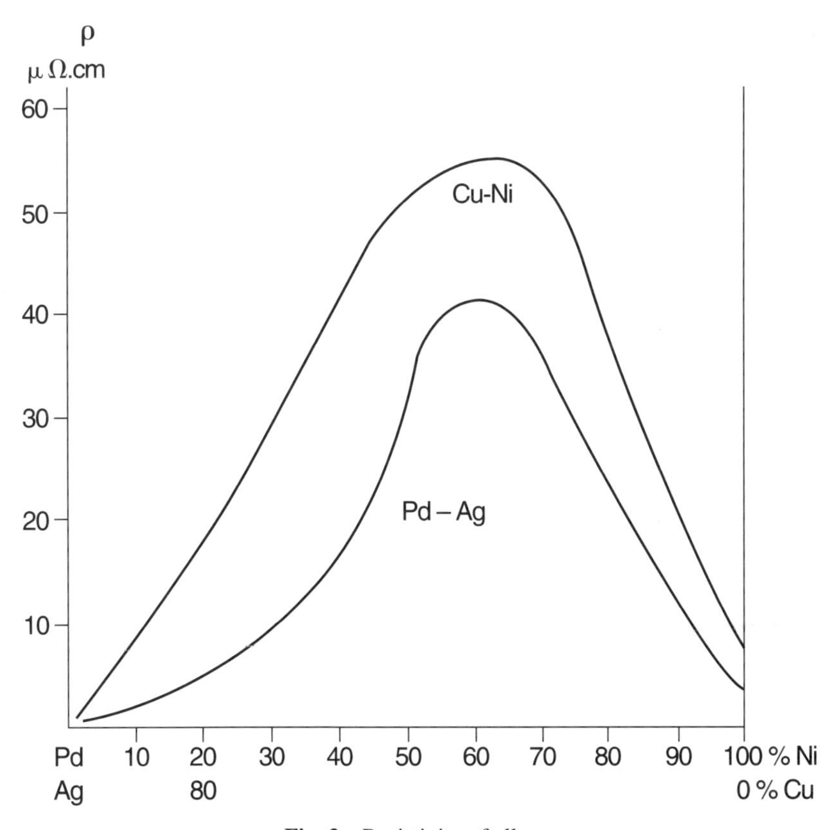

Fig. 3 Resistivity of alloys.

Solid solutions of Cu-Ni and Ni-Cr alloys help increase this temperature independence through distortion of the lattice. Moreover, the presence of Fe increases the probability of inelastic interactions, thus contributing to increased temperature-independent scattering. This effect is magnified in Ni-Cr through the presence of a Ni_3Cr super-lattice and short-range order (approximately 900–1100 nm) in the 540°C zone of the phase diagram. Experiments show that this phase, by a mechanism that is still only poorly understood, is partly responsible for the 10% greater resistivity at 300°K than at the γ phase, as well as for the low value of the TCR.

3. THIN FILMS

3.1 Thin Films in Use: Description and Electronic Properties

In the previous section we saw the usefulness of the electronic properties of the transition metals and their alloys for manufacturing resistors. These alloys are an obvious choice for the production of thin-film resistors.[10]

Two materials are commonly used in the manufacture of thin-film resistors: Ni-Cr and tantalum nitride.

Ni-Cr is obtained by vacuum evaporation or sputter deposition from a Ni-Cr alloy target. The composition of this alloy can vary between 60 and 80% (atomic composition) Ni. Experiments show that the deposit also contains oxygen, carbon, and impurities that are introduced by organic vapors in the vacuum systems and other miscellaneous contaminants from the surrounding enclosure. The electrical properties of these thin films *when they are deposited on 98% pure* Al_2O_3[11] *ceramics* are as follows:

Resistivity: varies between 200 and 400 $\mu\Omega \cdot$ cm
Sheet resistance:[12] between 15 Ω/■ and 300 Ω/■
TCR: between ±150 ppm/°C (depending on the conditions of deposition and any subsequent heat treatment)

[10]We refer to “thin films” as conductive, resistive, or dielectric materials that are deposited on substrates of different shapes and composition in the form of films whose thickness varies between 100 and 10,000 nm.

[11]Differences in the expansion coefficients of thin films and their substrates introduce variations (see Chap. 9 on precision resistors) in the resistivity and TCR of resistors, depending on the nature of the substrate. These variations can be calculated.

[12]Sheet resistance is expressed in Ω/square or Ω/■. This equation is calculated by taking a resistor of identical length and width L, and thickness e, which yields the

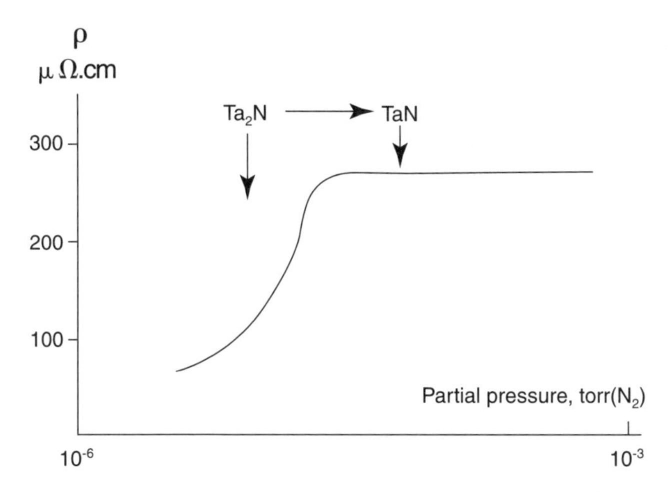

Fig. 4 Resistivity of deposited Ta-N vs. partial pressure of N_2.

Tantalum nitride is deposited by reactive sputtering in the presence of a plasma containing nitrogen.

In its pure, solid form, Ta is not a very attractive material for resistors. It has a resistivity of 13 $\mu\Omega \cdot$ cm with a TCR of approximately 4,000 ppm/°C. However, when it is deposited under the conditions indicated above, we increase the Ta_2N phase. This phase has the same crystalline structure as the β phase of thin-film tantalum, which is tetragonal.[13] But the structure and chemical composition of the nitrides of Ta are strongly influenced by the value of the partial pressure of nitrogen and the temperature of the substrate during deposition. The conditions for forming a stable thin film are obtained using partial pressures of nitrogen greater than 10^{-5} torr, as shown in Fig. 4,[14] as well as substrate temperatures above 180°C.

With a greater concentration of nitrogen, TaN can also be deposited. Under these conditions the resistivity of the material obtained, measured on the films located among the plateau of Fig. 4, is around 250–300 $\mu\Omega \cdot$ cm, with a TCR of approximately −80 to −100 ppm/°C.[15]

---

equation $R/\blacksquare = \rho L/eL = \rho/e$, where ρ and e are expressed in similar dimensions (for example, $\Omega \cdot$ cm and cm).

[13]The β phase is thermodynamically stable only in its thin-film form and does not exist as solid tantalum.

[14]Berry *et al., Thin-Film Technology.* [D. Van Nostrand], Princeton, 1968.
Couts, T. J. *Active and Passive Thin-Film Devices.* p. 91. Academic Press, 1978.

[15]The ultimate electronic properties of the films are obtained only after annealing at carefully determined temperatures for a predetermined period of time. The annealing conditions vary depending on the characteristics of the films, which are measured

Other materials are also used for thin-film resistors. These include Ni-P alloys obtained by chemical deposition (Kanigen metal) and the oxides of tin obtained by pyrolysis of tin chloride. These materials are principally employed in the manufacture of thin-film resistors on 96%-pure Al_2O_3 ceramic bars. The electrical performance of these resistors is generally far below that of resistors made with high-purity alumina substrates. However, they do have the advantage of operating at higher temperatures without damage.

Hafnium diborate is also used for high temperature applications such as the printer's thermal heads.

The new integrated passive electronic components (IPECs) are produced on silicon substrates such as semiconductor components. Besides the Ni-Cr and Ta_2N, materials like chromium silicon or titanium tungsten are used to build resistors on these IPECs.

3.2 Conduction Mechanisms in Thin Films and the Origins of Nonreversible Drift

Using Drude's approximation, electrical conductivity in a metal also can be applied to metallic thin films. In this case the film must be continuous; that is, its thickness must be high enough that the metal phases form a continuous medium without the presence of disconnected metallic clusters. However, when the thickness of the film is of the same order of magnitude as the mean free path of the carriers, the condition of isotropy on which Drude's treatment is based is no longer satisfied and the conditions at the limits of the Boltzmann transport equation (see Chap. 1) must account for it.

Although we will not discuss the development of their arguments, Fuchs[16] and Sondheim demonstrated that if σ_B is the conductivity of the solid metal, σ_F the conductivity of the thin film, and ζ the ratio between

immediately after deposition, and are considered an element of the manufacturer's technical expertise.

[16]Fuchs, K., *Proc. Cambridge Phil. Soc.* 34 (1938): 100.
Sondheim, F. H. *Phys. Rev.* 80 (1950): 401.
Sondheim, E. H., *Advan. Phys.* 1 (1952): 1.
See also the summaries in the following references:
Maissel, L. I. and Gland, R., *Handbook of Thin-Film Technology.* Chap. 13. McGraw-Hill, 1970.
Chopra, K. L., *Thin-Film Phenomena.* Chap. 6. Huntington, NY: Robert E. Krieger Publishing Co., 1979.

the thickness of the film d and the mean free path l of the carriers ($\zeta = d/l$), the ratio between σ_F and σ_B varies according to the following equations:

$$\frac{\sigma_F}{\sigma_B} = 1 - \frac{3}{8\zeta} \qquad \zeta \gg 1 \tag{1.2a}$$

$$\frac{\sigma_F}{\sigma_B} = \frac{3\zeta}{4} \cdot \left[\ln(1/\zeta) + 0.423 \right] \qquad \zeta \ll 1 \tag{1.2b}$$

If we assume that a fraction p of the carriers ($0 \leq p \leq 1$) is elastically diffused following a collision with the surface of the film, an impurity, or a grain boundary, Eqs. 1.2a and 1.2b become:

$$\frac{\sigma_F}{\sigma_B} = 1 - \frac{3(1-p)}{8\zeta} \qquad \zeta \gg 1 \tag{1.2c}$$

$$\frac{\sigma_F}{\sigma_B} = \frac{3\zeta(1+2p)}{4} \cdot \left[\ln(1/\zeta) + 0.423 \right] \qquad \zeta \ll 1 \tag{1.2d}$$

The ratio p can thus vary as a function of temperature and impurities or defects in the metal. Under these conditions, if α_F is the TCR of the thin film and α_B the TCR of the metal or solid alloy, the ratio α_F/α_B is governed by the following equations:

$$\frac{\alpha_F}{\alpha_B} = 1 - \frac{3(1-p)}{8\zeta} \qquad \zeta \gg 1 \tag{1.3a}$$

$$\frac{\alpha_F}{\alpha_B} = \frac{1}{ln(1/\zeta) + 0.423} \qquad \zeta \ll 1 \tag{1.3b}$$

Figure 5 represents the change in resistivity at 300°K of gold films deposited on mica as a function of the thickness of the film.

In practice, for thin films used to manufacture resistors, ζ will always be quite close to 1. This means that by systematically introducing impurities or crystal defects, changes in p can be used to create resistors that conform to the desired performance.

The thickness will also have an effect on the TCR curves and, in the case of Ni-Cr, these will be negative at low temperatures and positive at temperatures greater than 25°C.

The process of manufacturing thin films involves the bombardment of a substrate with the constituent atoms of the film. These coalesce on the substrate, creating films whose crystal structure is extremely disorderly

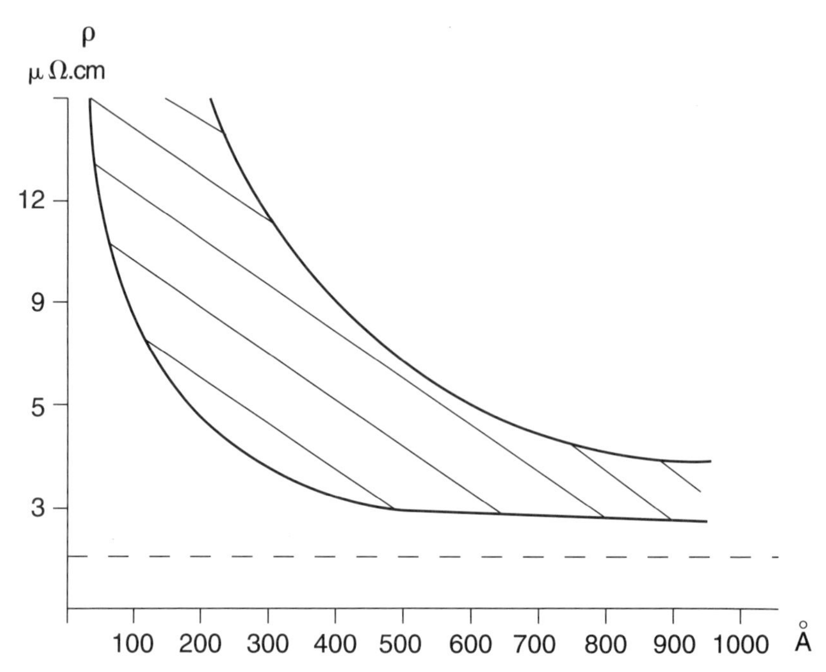

Fig. 5 Change in resistivity of a gold film deposited on mica as a function of film thickness. The dashed line represents the limit of resistivity of gold in its solid state.

and which possess an extremely high defect density (especially vacancies and interstitials). Through the τ_{imp} component (Matthiessen's rule), this results in an increase in resistivity compared to that of the metal and solid alloy. The irreversible drift that occurs during operation, which depends on temperature and time, can be controlled by activation phenomena that follow the Arrhenius law. In this way measurements have been made of the activation energy of a vacancy in gold deposits (0.54 eV).

In practice, crystal defects are not the only cause of such drift. The most drastic changes in resistivity over time are due to oxidation and corrosion phenomena caused by chemical reactions arising in the thin film from ionic components, humidity, and the electromigration of impurities and metal caused by the electric field.

The following chapters on thin-film resistors will provide a fuller analysis of these phenomena and demonstrate the various methods used to prevent them.

3.3 Substrates

Although the above remarks primarily concern thin-film components, we need to consider the nature of the substrate used in discussing these phenomena. It is in fact more appropriate to discuss the behavior of a "system" formed from a thin film and a substrate rather than an ideal thin film without substrate.

From this point of view the influence of the substrate on the electronic properties and behavior of the film should be as neutral as possible. But tests performed on films of Ni-Cr have shown that the results of irreversible drift as a function of temperature and time were significantly improved when using alumina and silicon substrates rather than glass. There are a number of factors that we need to take into account when selecting a substrate, depending on the desired performance of the "system." Table 1 summarizes the selection criteria of such substrates as a function of the thin-film criteria selected.

We see, therefore, that the substrate must satisfy a number of criteria in order to obtain high-performance thin-film devices. It is worth repeating that, despite the inherent physical qualities of the film and its environment (contacts, etc.), it is the thin film-substrate unit that is principally responsible for the final performance of thin-film resistors.

The most commonly used substrates are porcelain, alumina-based ceramics between 96% and 99% pure, sapphire, and surface oxidized silicon.

TABLE 1

Criteria of Substrate Selection	Results with Thin Films
Smooth atomic surface	Film of uniform density
Clean surface	Good adhesion
Good flatness	Well defined geometry
Nonporous	No outgassing
Coefficient of expansion α_s	No limitations if α_s is almost equal to the α_F of the film
Thermal conductivity	Minimize high T° gradients
Thermal stability	Minimize thermal deformations
Chemical stability	No contamination
High insulation resistance	No insulation defects
Low cost	Obvious benefit for industrial applications

The best of them (99% Al_2O_3, sapphire, and silicon) are used for integrated networks of precision resistors.

4. THICK- AND THIN-FILM CERMETS

4.1 Conductivity Mechanism

Cermet is the generic name for a group of materials formed from mixtures of conducting and insulating materials (cermet = ceramic plus metal). Physics treats cermets as solid composites formed from mixtures of different materials. In this sense the electrical properties of cermets can be explained and predicted using topological methods based on the theory of *percolation.* This theory will also be used to account for the properties of certain amorphous thin films.

The percolation models for cermets are based on the composite structures of conducting and insulating grains. The theory consists of determining how a series of conducting sites (conducting grains) that are randomly distributed in a two- or three-dimensional space among insulating sites (insulating grains) are interconnected, and in measuring the resistivity of the material as a function of the relative concentration of conducting and insulating grains.

One of the models illustrating this concept is supplied by a random mixture of conducting and nonconducting[17] balls of the same radius. We use a container closed on the top and bottom, and filled with a certain ratio p of copper-plated balls and 1-p insulating balls. The change in conductivity (Fig. 6) is represented by a curve of the following type:

$$\sigma(p) \cong \sigma_0(p - p_c)^{\beta} \tag{1.4}$$

in which p_c is a critical threshold of concentration beyond which there is no conducting path. The value of the threshold ($p_c \cong 0.27$) is a universal constant present in all percolation phenomena, σ_0 is the value of the conductivity of a single path, and β is equal to 2 near the threshold p_c but varies with the nature, topology, and concentration of the conducting material. Onto this model are superimposed the effects of the position disorder of the balls (which make direct calculation of the conductivity of the recipient when filled with conducting balls impossible) and of the *composition* of

[17]"Applications de la percolation." Groupe de Physique des systèmes désordonnés, *Ann. de Physique,* 1982.

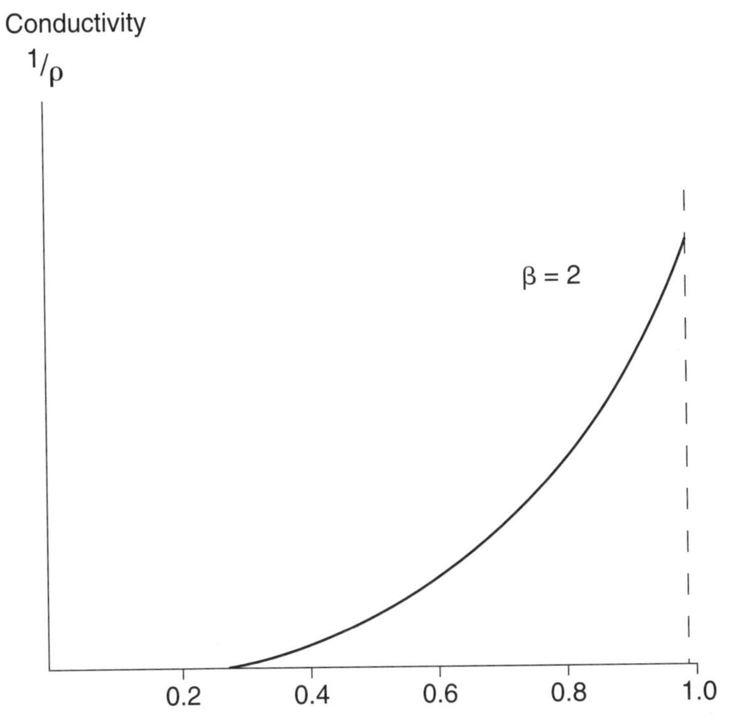

Fig. 6 The curve above represents the variation in conductivity of the mixture as a function of the ratio *p*, where p = number of conducting balls/total number of balls.

the material, represented by the random presence of conducting or insulating objects. This last type of disorder is the only type that is important when evaluating the critical behavior of the system, represented by β.[18] Figure 6 represents this phenomenon graphically.

In practice, the model is complicated because of differences in the size and inherent density of the insulators and conductors in the mixture. This means that $\beta = 1.35$–1.50 and p_c varies between 0.10 and 0.27.[19]

Theoretical and experimental studies of conductivity of such media in alternating current show that the critical threshold varies with frequency. Interpretation of these phenomena also shows that the simplified

[18]We can also study the behavior of mixtures of this type by examining their fractal dimensions, which lie between 2 and 3.

[19]Clerc, J. P., "Effet de taille et d'anisotropie en percolation en relation avec la conductivité." Thesis. University of Provence, 1980.

percolation model does not take into account all the effects of conductivity. This model must be accompanied by a more complex model of a random network of conducting threads and inter-grain capacities. This model can be used to explain both the frequency response and the noise phenomena that are specific to these materials.

Figure 7 represents the conductivity of the glass-Ag cermet as a function of the silver concentration:

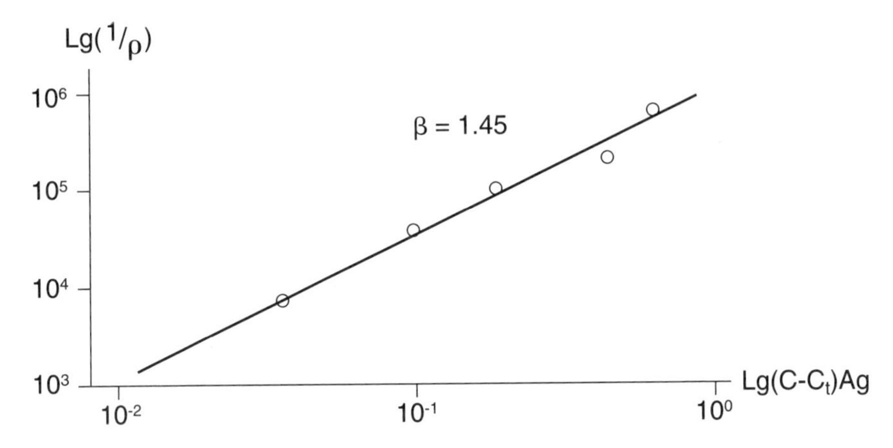

Fig. 7 The x axis represents the Napierian log of the difference between the concentration of Ag and the threshold concentration. The y axis represents the log of conductivity.

We see that in this case $p_c = 0.1$ and $\beta = 1.45$.

A mixture of diallyl-phthalate and carbon black, for example, will give $p_c = 0.25$ and $\beta = 1.38$. Despite the considerable differences in the nature of the materials used in the first example of Fig. 6, the orders of magnitude for p_c and β are the same, which supports the model.

4.2 Thick-Film Glass-Metal Cermets

The electrical and mechanical properties of these cermets will be described in greater detail in the following chapters, which discuss their use in the field of electronics. Here we will describe some of their individual properties to illustrate the physical phenomena and mechanisms described above. Their manufacture requires high temperatures, and they are used in resistors and conductors. They consist of mixtures of glass beads (whose diameter is between 0.5 and several microns) with a low melting point and free of ions (borosilicates), and of fine metal, alloy, or oxide particles (their size is between several nanometers and a few tens of nanometers). Organic products (terpenes, resins, etc.) with low evaporation points are added to the mixture to control its rheology. Inks are deposited on ceramic substrates (Al_2O_3 or any other ceramic material) by silk screening, using masks to represent the circuit geometry. The cermet is heat treated at low temperature (T° < 600°K) to eliminate the organic binder. The temperature is then raised to the melting point of glass (T° between 1000 and 1400 °K) so that the material coalesces. The thickness of the film after heat treatment is approximately 20–25 microns.

4.2.1 Conductive cermets Conductive cermets are made from mixtures of glass and metal. They are characterized by a resistivity of approximately 25–40 mΩ/■. Their TCRs are generally smaller by several ten percents than those of the solid metals used in their composition. The following metals are used:

Silver: This mixture is still used for hybrid film applications. Its resistivity is around 25–35 mΩ/■. Unfortunately, silver migrates easily under the influence of an electric field and humidity (electromigration).

Gold: This mixture is used when semiconductor chips require the presence of gold-silicon eutectic connections. The cost of this cermet is high, however, and its solderability poor.

Gold-Platinum and Palladium-Gold: These mixtures are the most commonly used in hybrid films and for the conducting elements in resistors. Migration is minimal and solderability very good.

4.2.2 Resistive cermets The "resistive" cermets are characterized by a resistivity between 100 and 10^4 Ω/■. One of the difficulties in developing this type of cermet is caused by the variations in the threshold value p_c. For this reason, and to reduce the scattering of ohmic values, the metal is replaced with oxides or complex mixtures. The values of the coefficients of resistance as a function of temperature (TCR) and voltage (VCR) reflect the semiconducting properties specific to these materials, in which conductivity phenomena arc very sensitive to changes in the electric field, temperature, and the glass-oxide interface. Thus, the TCR and VCR are generally negative and range between −100–500 ppm/°C for TCR and −10–30 ppm/volt for resistors in which the length/width ratio is of the order of two. The following materials are used in their construction: *oxides of thallium, rhenium, ruthenium, etc.* The resistivity of these materials, which are mixed with lead-borosilicate glasses, when sintered on ceramic substrates, ranges between 50 Ω/■ and 10^6 Ω/■ for thicknesses of 20–25 μm.

4.3 Thick-Film Organic Cermets

Cermets of this type are used in consumer goods, in which the cost of ceramic substrates is prohibitive, or in potentiometric movement sensors, because of their resistance to wear.

In the case of conducting cermets, the glass binder is replaced by various polymers of the diallyl-phthalate type. The composition and the dopants used are generally kept confidential by the manufacturers of these materials.

In resistive cermets the various oxides are replaced with carbon black embedded in a polymer matrix. Under these conditions, although Ohm's law still holds for this material, the conductivity is not linear with the field, even as a first approximation (it is only linear within small variations of the electric field). This is caused by the specific mechanisms of electrical conduction among the particles of carbon black. Conduction is highly dependent on the interfaces and takes place by tunneling between particles, the probability of an electron jump being highly dependent on the field and the composition of the interface.[20] The overall effect of conductivity is also highly sensitive to contact resistances between the material, and the current and voltage electrodes. This resistance is generally not ohmic, and a typical space charge effect at the conductor-cermet interface is produced. Figure 8 shows the nonlinearity of Eq. 1.4 for cermets of this type. Because of this, it is practically impossible to predict the resistivity and behavior of these materials based on their inherent characteristics. The only reliable measurements are those made on components manufactured from these cermets.

The TCR of these materials is always negative (approximately −1000–5000 ppm/°C) as is the VCR (on the order of −0.1%/volt).

4.4 Thin-Film Cermets

Thin-film cermets are heterogeneous mixtures of conductive and dielectric materials of varying types and structures. The difference with thick-film cermets resides in their thickness and methods of production: vacuum deposition or sputtering in a reactive plasma. As with thick-film

[20]Enid Keil Sichel., *Carbon Black Polymer Composites.* Marcel Dekker Inc., 1982.

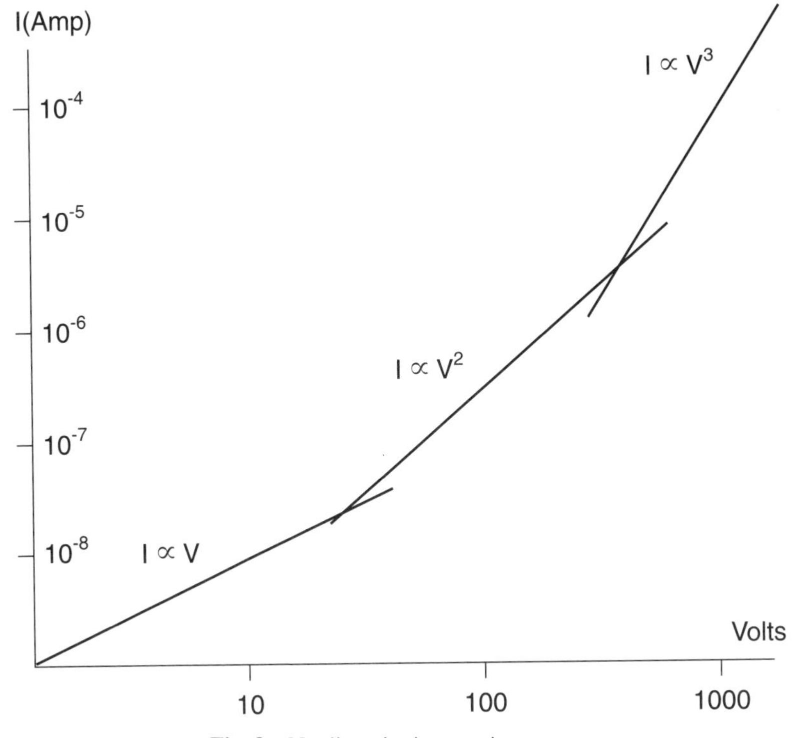

Fig. 8 Nonlinearity in organic cermets.

cermets, the conductivity of thin-film cermets is very sensitive to even minor changes in relative composition and structure, and the theory of percolation is the tool of choice in analyzing their electronic properties. The important factor in resistors using this type of structure is not the resistivity of the conducting phase, but the way in which the various crystallites are interconnected.

Using this approximation, it has been shown[21] that the conductivity is governed by tunnel-effect phenomena among the crystallites, as well as by diffusion at the grain boundaries of crystal clusters.

Mayadas and Shatzkes[22] have shown that when the mean free path λ_0 (λ_0 includes *all* diffusion mechanisms, impurities, defects, etc.) is smaller than the average diameter of the grains in a polycrystalline thin film, which is the case with thin-film cermets, diffusion at the grain boundaries and their orientation

[21]Chopra, K. L., *Thin Film Phenomena.* Robert E. Krieger Publishing, 1979.

[22]Mayadas, A. F. and Shatzkes, M., *Phys. Rev.* B1 (1970): 1982.

with respect to the electric field are among the principal factors governing conductivity. The size and orientation of the grains are controlled by the conditions under which the film has been deposited, such as rate of deposition, the energy of the atoms reaching the substrate, temperature of the substrate, etc. The size of the grain and its related effects thus depend on the methods of preparation and are not inherent properties of the general structure of thin films.

If we apply the assumption that the grain boundaries are defined by Dirac functions, we can apply the Boltzmann transport equation under these conditions and calculate an overall relaxation time for all the diffusion mechanisms by using Matthiessen's[23] rule. Mayadas and Shatzkes also obtained a relaxation time τ_{grains} as a function of the parameter α_0, which describes the geometry of the grain and its collision cross-section:

$$\alpha_0 = \frac{\lambda_B}{\langle r \rangle} \cdot \frac{D}{1 - D} \tag{1.5}$$

in which λ_0 is the mean free path of the solid material, $\langle r \rangle$ the average grain diameter and D the coefficient of diffusion of the grain boundary. The ratio of the conductivity of the thin film to conductivity of the solid material is given by the equation:

$$\frac{\sigma_F}{\sigma_B} = f(\alpha_0) = 1 - \frac{3\alpha_0}{2} + 3\alpha_0^2 - 3\alpha_0^3 \, ln(1 + 1/\alpha_0) \tag{1.6}$$

in which σ_F is the conductivity of the thin film and σ_B is the conductivity of the solid material.

In addition, Mayadas and Shatzkes have shown that the grain size, again by means of Matthiessen's rule, also controls the material's TCR. We can thus estimate that the TCR of an Ni-Cr thin film containing Cr (Cr_2O_3) oxides is very close to zero for $\langle r \rangle \cong 50$ nm, which has been experimentally verified. The other conduction mechanism is a tunneling effect between conducting clusters. The conductivity is proportional to

$$\sigma = \sigma_0 \exp[-(\delta E/kT)^{n}] \tag{1.7}$$

in which δE is the electrical field between the conducting clusters and n ($\cong 1/2$) varies with the nature of the dielectric[24] at the boundary of the

[23]Matthiessen's rule states that the scattering mechanisms for the charge carriers responsible for conductivity act independently of one another. This rule is covered in greater detail in Chap. 3.

[24]A detailed discussion of the mechanisms of field effect conduction is beyond the scope of the present work. Interested readers may wish to consult K. L. Chopra, *Thin Film Phenomena,* Robert Krieger Publishing Company, 1979, along with T. J. Coutts, *Active and Passive Thin Film Devices,* Academic Press, 1978.

cluster. The TCR of such a mechanism will vary according to the following relation:[25]

$$\mathrm{TCR} \cong \delta E \cdot (kT)^{1/2} \cot[2\pi\delta E \cdot (kT)^{3/2}] - \frac{1}{T} \tag{1.8}$$

In the current model, the TCR is inversely proportional to temperature and the TCR of thin-metal films is positive. Consequently, by means of Matthiesen's rule, this effect is often used to obtain very low TCRs, often below ±5 ppm/°C, through controlled oxidation of thin-film metal alloys.

Table 2 lists the principal thin-film cermets with their electronic characteristics.

The behavior of cermets subject to irreversible drift due to the effects of temperature and time, or of an electric field, is much more difficult to determine than in the case of conventional thin films.

Although the antagonistic phenomena discussed in Chap. 2 for the most part follow the Arrhenius law and thus have activation energies, they act simultaneously and the final effect results in a semi-empirical model, as shown by the following equation:

$$\Delta R/R = Ct \cdot t^n \cdot \exp\left[-\sum \frac{\Delta G^0}{kT}\right] \tag{1.9}$$

TABLE 2

Film Material	Resistance in Ω/■	TCR in ppm/°C
NiCr (50:40 at.) CrO_2 O_2 (7:3 at.)[a]	20–200	±5
Cr-Si-O_2 (24:40:36 at.)[b]	100–4000	±250
Cr-Ti (35:65 at.)[c]	250–650	±150
Cr-SiO (70:30 at.)[d]	600	−100 to −1000
Cr-CrO_2 [3] (50:50 at.)[e]	10–1000	−500
SnO_2[f]	3–10	−1000
TaN-TaN_2[g]	100–500	−200 to −25

[a]Personal communication from Sfernice Research Department.
[b]Personal communication from Sfernice Research Department.
[c]E. R. Layer and H. R. Olson, *Elec. Mfg.* v. 58, p. 104 (1956).
[d]W. O. Freitag and V. R. Weiss, *Res. Dev.*, p. 44, August 1967.
[e]Personal communication from Sfernice Research Department.
[f]L. Bolland and G. Siddal, *Vacuum,* v. 3, p. 375 (1955).
[g]Personal communication from Sfernice Research Department.

[25]T. J. Coutts, *op. cit.*, pp. 165–204.

This model includes phenomena involving the diffusion of oxygen at grain boundaries or on the surface, grain growth, defect healing, and oxidation. Equation 1.9, in which n ranges between 1/2 and 3/2, has often been verified during aging measurements of thin films as a function of temperature.

In general, specialists will use any physico-chemical means available to obtain energies of activation that are as high as possible so as to minimize irreversible drift. For example, Figs. 9–11 represent changes in resistivity as a function of temperature and time for Ni-Cr-Si and Ni-Cr-Cr_2O_3 cermets.

These figures illustrate the phenomena described in Chap. 2. In Fig. 9, the aging curve at 430°C is typical of the healing of vacancy defects, interstitials, and dislocations. After an increase in resistivity, likely due to surface oxidation ($Cr \Rightarrow Cr_2O_3$), the oxide layer thus formed acts as a protective barrier of depth z [oxidation follows an $A \cdot erf\,(z/(At)^{1/2})$ law, refer to Chap. 2]. The resistivity then decreases sharply as a result of the high crystal defects). This is corroborated by the sudden change in TCR (Fig. 10), which goes from a negative to a strongly positive value.

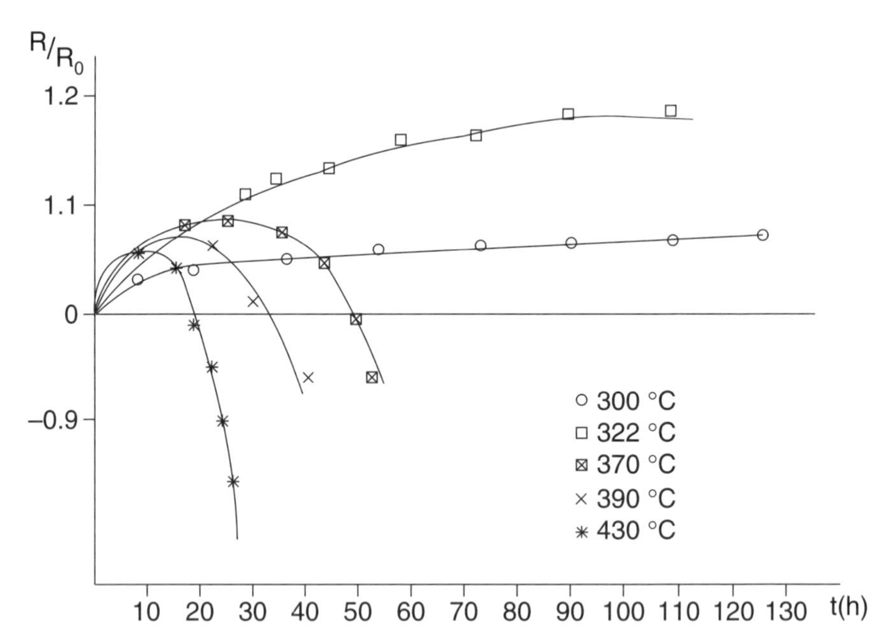

Fig. 9 Variations in the value of a normalized resistance of a sputter-deposited Ni-Cr-Si film as a function of time and temperature.

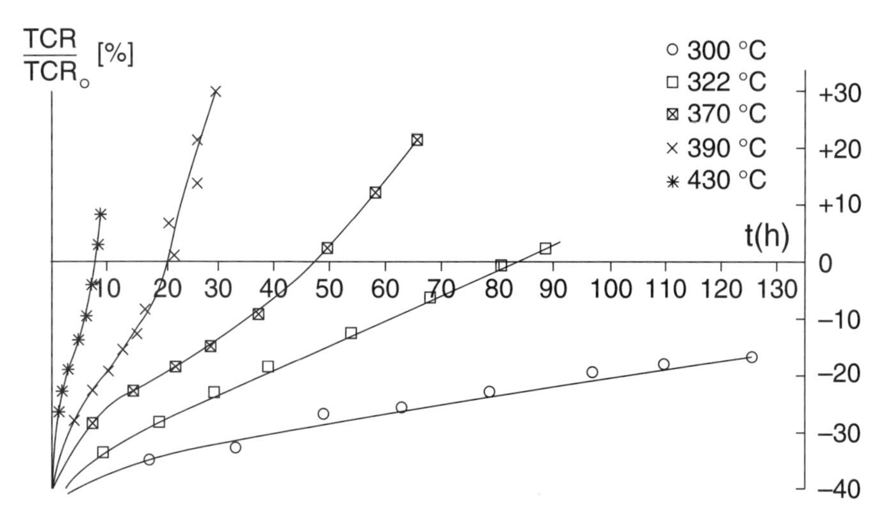

Fig. 10 Variation in TCR of the film shown in Fig. 9 as a function of the same temperatures and time.

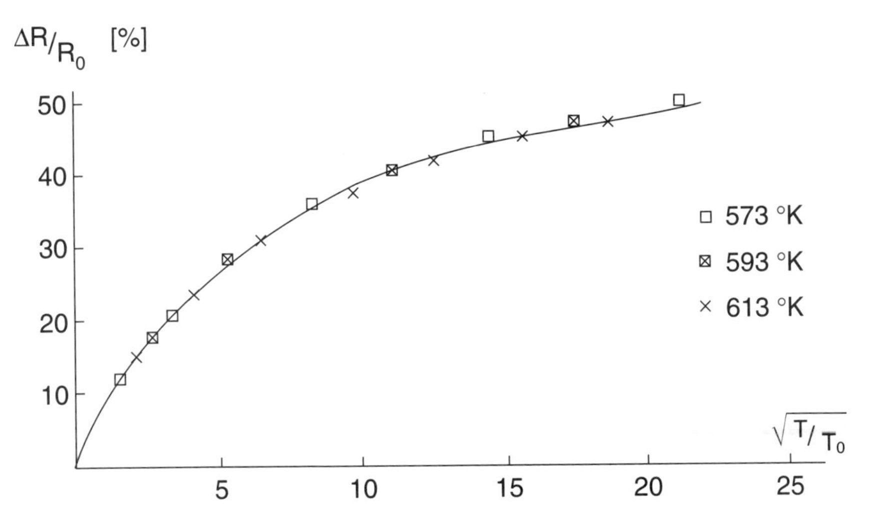

Fig. 11 Normalized drift curve for a Ni-Cr-Cr_2O_3 film at various temperatures with respect to a reference $T_0 \cdot T_0 = 573°K$, $t_0 = 1$ hour, the activation energy of aging has a value of 2.1 eV, and $\lambda_0 \cong 50$ nm.

The drop in processing temperature gradually reduces the rate of defect healing, as shown by the aging curves at 390°C, 370°C, 322°C, and 300°C, in such a way that the process is progressively governed by oxidation phenomena. This is corroborated by the reduced growth of the TCR.

Figure 11, which illustrates the irreversible long-term drift of a Ni-Cr-Cr_2O_3 cermet, confirms the fact that, for low temperatures and after the initial suppression of crystal defects, aging is controlled by oxidation phenomena, which increase the resistivity of the film.

5. CONCLUSION

The materials used to produce resistors require complex manufacturing procedures, which are certainly comparable to the techniques used in the semiconductor industry. In particular, the need to obtain materials with high resistance to mechanical, electrical, climatic, and thermal wear, together with TCR near 0 ppm/°C, has required very close collaboration among metallurgical research laboratories, alloy developers, and resistor manufacturers.

5

Noise Phenomena

1. INTRODUCTION

In Chap. 1 we showed that current density in a resistive or conductive material is proportional to the applied electric field if its value is less than the ionization threshold of the atoms in the material. Chapters 2 and 3 discussed the reversible and irreversible drifts occurring in these materials.

There is, however, another phenomenon, which invalidates the proportionality relationship between electric field and current. In fact, experiments show that the current crossing a resistive material or resistance contains a random component that modulates the average current. This random component, which degrades the performance of the resistor, is noise. Chapter 5 covers the mechanisms and sources of noise in resistors and the methods used to measure and minimize them.

2. ORIGINS OF NOISE

Noise in electronic components arises from sources that are part of the components or the system itself, or from external sources. An external source of noise is referred to as *interference*. The sources of interference, whose origins are extremely varied, can be identified and eliminated through shielding or other methods. This is not true, however, for sources within the component. We have seen in Chap. 2 that when the temperature in a material is different from 0°K, fluctuations occur in the average distance among atoms in that material. This is also true for the charge carriers (electrons in the majority of phenomena involving resistors) responsible for current flow in the material. This movement is the primary cause of noise, which can be broken down into two main sources: thermal noise and current noise, the latter being often referred to as excess noise or 1/f noise.

2.1 Thermal Noise

In Chap. 2 we saw that in the case of diffusion phenomena that do not involve a transport force, the mean displacement of an atom from its point of origin is zero. The linear displacement, however, is not zero and possesses a variance equal to:

$$\langle L(t)^2 \rangle = \nu \lambda^2 t = 2Dt = 2\tilde{M}kTt \qquad (1.1)$$

in which λ is the interplanar spacing, ν the jump frequency among crystal sites, D the coefficient of diffusion, and $\tilde{M}$ the mobility of the components of the system in the thermal environment with respect to the length $L(t)$. This equation is considered to be absolutely universal, as was shown by Einstein,[1] and it can be applied to any observable parameter. Then, it can be applied to charge carriers or electrons in a crystal lattice subject to thermal fluctuations. In this case:

$$\langle \delta q^2 \rangle = 2GkTt \qquad (1.2)$$

in which G is the conductivity and $G = 1/R$ (dependent on the mobility). In this way, in a 1-ohm resistor ($G = 1$ ohm^{-1}) at a temperature of 300°K, measurement of the charge fluctuation at the ends of the resistor is approximately $\langle \delta q^2 \rangle^{1/2} \cong 10^{-10}$ coulombs per second. In other words the resistor displays a current equivalent to 10^{-10} amperes. Obviously this could also be applied to the average displacement speed of the carriers subject to the action of an electric field. Shortly after Einstein's work, Langevin[2] analyzed Brownian movement by considering, as Drude had done for Ohm's law, the restoring forces with a relaxation time τ. His work is a model of its type and a pleasure to read, and culminates with the equation that is commonly referred to as Nyquist's or Johnson's formula, or as white noise,

[1]Einstein, A., *Ann. Phys.* 17 (1905): 549 and *Ann. Phys.* 19 (1906): 289, 371.

[2]Langevin, M. P., *Comptes Rendus Acad. Sci.* 140 (1908): 530.

because the spectral density of the voltage is a function that is independent of the frequency for all octaves:

$$<V_f^2> = 4RkTdf \qquad (1.3)^3$$

in which T is the temperature of the resistor in °K, R the value of the resistance in Ω, k Boltzmann's constant ($1.38 \cdot 10^{-23}\ J \cdot K^{\circ -1}$), and df the range of a frequency band of arbitrary width.

This equation is equivalent to Einstein's formula above. It shows that thermal noise increases with temperature, frequency, and resistance. Its value is not insignificant. For example, a 100 kΩ resistor at a temperature of 300°K has a root-mean-square (rms) noise of 4 μV when measured with an instrument that has a 10 kHz bandwidth. When the resistor is used as an input impedance for an amplifier system in which the gain is a function $A(f)$ of frequency, the response due to white noise is equal to:

$$<V^2> = 4kT \int_0^{\infty} R(f)|A(f)|^2 df \qquad (1.4)$$

in which $R(f)$ is also dependent on frequency.

2.2 Excess Noise, Current Fluctuations

If $N(t)$ represents the random number of charge carriers present in a volume V of a conductor and if $v_i(t)$ is the velocity along the axis on which a current of charge q associated with $N(t)$ is measured, then:

$$I(t) = \frac{q}{L} \sum_{1}^{N(t)} v_i(t) \qquad (1.5)$$

Using this equation it can be shown[4] that the variance σ_1^2 of the current follows an equation:

$$\sigma_1^2 = \frac{q^2}{L^2} (\sigma_N^2 <v>^2 + \sigma_v^2 <N>) \qquad (1.6)$$

in which σ_N^2 is the variance associated with the fluctuation in the number of carriers and σ_v^2 is the variance associated with the fluctuation in velocity.

[3]The correct form of the Langevin equation is $<v_f^2> = \dfrac{4RkTdf}{1 - (2\pi f\tau)^2}$ in which $\tau = L/R$, L being the self-inductance of the resistance.

[4]Pinet, D., "Le Bruit de fond en l/f comme critère de qualité des résistances en couches minces." Thesis. University of Montpellier, 1987.

We can estimate the size of σ_1^2 to determine the excess noise.

When the average current is not zero, the velocity vector of carriers is formed from two components: thermal velocity $\vec{v}_T$ and $\vec{v}_D$, the drift velocity caused by electric field $\vec{E}$. Then $\vec{v} = \vec{v}_T + \vec{v}_D$.

With each isotropic diffusion, the thermal velocity with absolute value $\|\vec{v}_T|$ is equally distributed over a sphere of radius kT/m (m being the mass of the carrier). After each diffusion, the velocity $|\vec{v}_D|$, which is always parallel to the field, assumes a value dependent on the acceleration of the electric field: $v_D = qE/m + v_c$, the constant v_c depending on the structure of the material's conduction bands. As an example, $\vec{v}_c = 0$ in a covalent semiconductor and $\vec{v}_c = -a\vec{v}_D$ in polar conductors, where a is a constant depending on the conductor material. We can express this in terms of the variance, as follows:

$$\sigma_v^2 = \sigma_{v_T}^2 + \sigma_{v_D}^2 \tag{1.7}$$

When the average current is zero, $\sigma_v^2 = \sigma_{v_T}^2$, and the white noise described in the previous paragraph reappears. That is:

$$\sigma_{v_T}^2 = \frac{kT}{m} \quad \text{and} \quad \sigma_{I_T}^2 = \frac{kGT}{\tau} \tag{1.8}$$

in which τ is the relaxation time between each collision of a charge carrier.

When the average current is not zero, $\sigma_{V_D}^2$ can easily be calculated based on the following diagram, in which v_D still follows the direction of the current. Figure 1 represents the shape of v_D.

Calculation of the mean and mean square of $v_D - v_c$ yields the following:[5]

$$\langle v_D - v_c \rangle = \frac{qE}{2m} \cdot \frac{\langle t^2 \rangle}{\langle t \rangle} \quad \text{and} \quad \langle (v_D - v_c)^2 \rangle = \frac{q^2E^2}{3m^2} \cdot \frac{\langle t^3 \rangle}{\langle t \rangle} \tag{1.9}$$

consequently,

$$\sigma_{V_D}^2 = \langle (v_D - v_c)^2 \rangle - \langle (v_D - v_c) \rangle^2 = \frac{q^2E^2}{m^2}\left[\frac{\langle t^3 \rangle}{3\langle t \rangle} - \left(\frac{\langle t^2 \rangle}{2\langle t \rangle}\right)^2\right] \tag{1.10}$$

[5]Pinet, D., "Le Bruit de fond en 1/f comme critère de qualité des résistances couches minces." Thesis. University of Montpellier, 1987.

Savelli, M. *et al. l/f Noise as a Quality Criterion for Electronics Devices and Test Measurements.* Association Technique de l'Electronique, France, 1984.

Pinet, D. *et al.*, "l/f Noise in Thin-Film Resistors." Ninth International Conference on Noise in Physical Systems. Montreal, May 1987.

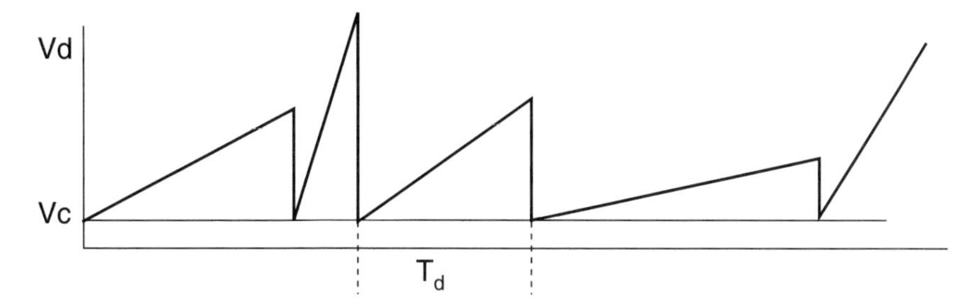

Fig. 1 Change in drift velocity v_D over time.

(which is the classical statistics theorem). Equation 1.5 can then be written:

$$\sigma_v^2 = \frac{kT}{m} + \frac{q^2E^2}{m^2} \cdot \left[\frac{\langle t^3 \rangle}{3\langle t \rangle} - \left(\frac{\langle t^2 \rangle}{2\langle t \rangle} \right)^2 \right] \tag{1.11}$$

Since the mean current density in a resistor of length L is equal to $\langle I \rangle = \langle N \rangle \langle v \rangle q/L$, then the mean current becomes:

$$\langle I \rangle = \frac{q}{L} \langle N \rangle \cdot \left[\frac{qE}{2m} \cdot \frac{\langle t^2 \rangle}{\langle t \rangle} + v_c \right]$$

from which:

$$\frac{qE}{m} = \left[\frac{\langle I \rangle L}{q \langle N \rangle} - v_c \right] \frac{2\langle t \rangle}{\langle t^2 \rangle} \tag{1.12}$$

Equation 1.11 can now be written:

$$\sigma_V^2 = \frac{kT}{m} + \left[\frac{\langle I \rangle L}{q \langle N \rangle} - v_c \right]^2 F(t) \tag{1.13}$$

in which

$$F(t) = \frac{4/3 \langle t^3 \rangle \langle t \rangle}{\langle t^2 \rangle^2} - 1$$

If N does not fluctuate, the current variance is expressed by the equation $\langle I^2 \rangle = q^2 \langle N \rangle \langle v^2 \rangle / L^2$. In general we can say that:

$$\sigma_1^2 = \frac{kGT}{\tau} + \frac{1}{\langle N \rangle} [\langle I \rangle - I_c]^2 \cdot F(t) \tag{1.14}$$

where $I_c = q/L \langle N \rangle v_c$

In semiconductor materials and other covalent materials, I_c is zero; however, it is different from zero in non-polar conductors, as quoted above. White noise is again present, along with a noise component that depends on the mean value of carrier relaxation and the number N of carriers per unit volume. When the distribution of t follows Poisson statistics, it can be shown that $F(t)$ equals one[6] and the spectral density of current noise is also independent of frequency. This would be useful if we assume that carrier diffusions are isotropic. But we have shown that this is not the case, at least for crystal defects and ions contained in the material. The distribution cannot, therefore, be Poissonian, and it has been shown[7] that, under constant conditions, $F(t)$ has the form $1/f$.

2.3 Noise Measurements

Figure 2 is an equivalent circuit diagram of noise in a test resistor R and its load resistor R_L, both with and without polarization [resistors R_L are wire-wound resistors with little (or no) excess noise compared to R]. S_0 represents the spectral density at the terminals of the two nonpolarized resistors and I_R and I_{RL} the two equivalent thermal noise generators. S_0 then equals:

$$S_0 = 4kTR_{\text{equ}} \qquad \text{with} \qquad R_{\text{equ}} = \frac{R_L \cdot R}{R_L + R}$$

When there is polarization the diagram is the same and an excess noise generator is added to the terminals of R. S_1 represents the total noise spectral voltage density.

$$S_1 = 4kTR_{\text{equ}} + S_{Ie}R_{\text{equ}}^2$$

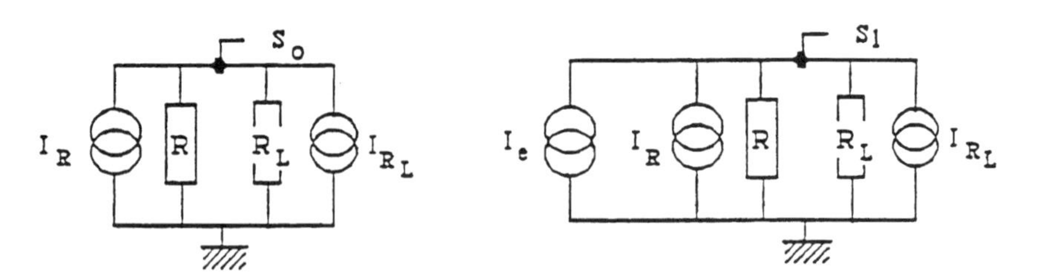

Fig. 2 Equivalent circuit diagram of noise—with and without polarization.

[6]If a series of collisions is Poissonian with a density $p = \tau - 1$, the interval between two collisions has a probability density of $p(t) = (1/\tau)e^{-t/\tau}$ and $\langle t \rangle = \tau$, $\langle t^2 \rangle = 2\tau^2$ and $\langle t^3 \rangle = 6\tau^3$, from which $F(t) = 1$.

[7]Macdonald, D. K. C., *Noise and Fluctuations*. Wiley, 1962.

in which S_{Ie} is the excess noise spectral current density. As a result:

$$S_1 - S_0 = S_{Ie} \cdot R_{\text{equ}}^2 \tag{1.15}$$

We use a dimensionless coefficient K to represent the excess noise spectral current density of a resistor Rn in which:

$$S_{Ie}(f) = \frac{kI^2}{f} \quad (A^2/Hz) \qquad \text{or} \qquad S_e(f) = S_{Ie}(f)R^2 \quad (V^2/Hz) \tag{1.16}$$

Measurements are expressed in the form of an excess noise index in the frequency band $f_2 - f_1$, such that:

$$\delta V_e^2 = \int_{f_1}^{f_2} KI^2R^2 \frac{df}{f} = KI^2R^2 \cdot \ln\left[\frac{f_2}{f_1}\right] \tag{1.17}$$

When the band extends across a decade of frequencies, we have:

$$\delta V_{e/d}^2 = KI^2R^2 \ln(10) \tag{1.18}$$

The ratio $\delta V_{e/d}^2$ expressed as rms μvolt by the dc voltage in volts, gives the noise index NI:

$$\text{NI} = \frac{\delta V_{e/d}^2}{V} = (2 \cdot 3K)^{1/2} \tag{1.19}$$

which is often expressed in decibels:

$$(\text{NI})_{\text{dB}} = 20\ln\left(\delta V_{e/d}^2/V\right) \tag{1.20}$$

Figure 3a represents the excess noise spectral density of a Ni-Cr thin-layer resistor of 10 kΩ as a function of the current I. Equation 1.16 is well confirmed for higher currents. For weaker currents, the spectral density is not proportional to I^2 but is nonlinear with respect to I^2. Figure 3b, however, shows that the $1/f$ law is obeyed fairly well.

A cut-off frequency exists beyond which white noise merges with $1/f$ noise. This frequency is given by the equation:

$$K \cdot \frac{I^2R^2}{f_c} = 4kRT, \quad \text{that is } f_c = K \cdot \frac{I^2R}{4kT} \tag{1.21}$$

Experiments show that the value of K varies as a function of the material, the quality of the contacts, and any defects in the resistor, as well as its geometry. There is excellent correlation[8] between the value of $\langle \delta V^2 \rangle$ at very low frequencies (less than 9 hertz) and irreversible drift.

[8]See Pinet, *op. cit.*

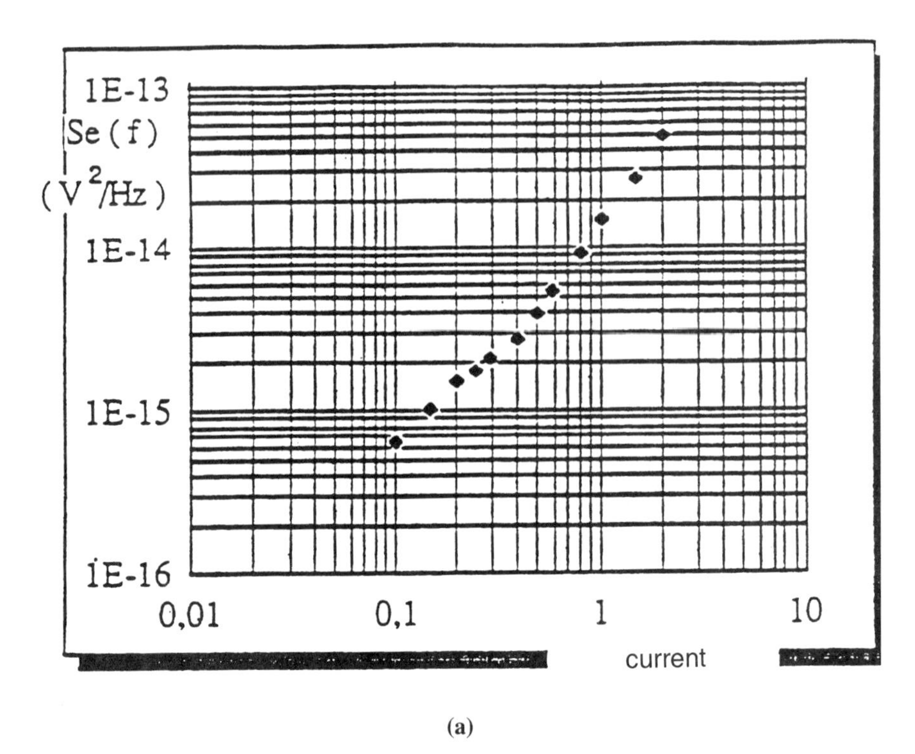

(a)

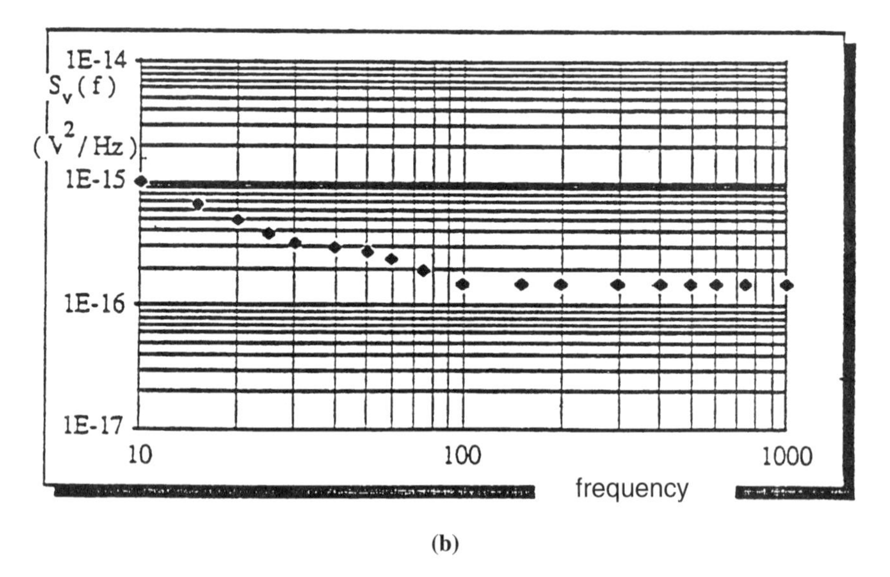

(b)

Fig. 3 Spectral density of excess noise.

Table 1 lists the orders of magnitude of K as a function of the technologies used to manufacture the resistors. We see that in all cases, except for thick-layer cermets, noise levels are extremely low and very close to one another.

Here, E_0 represents the noise voltage of the two parallel resistors unpolarized in nanoV/Hz$^{1/2}$, and E_1 represents the noise voltage at the terminals of the two resistors crossed by a current expressed in mA. The index NI is given in decibels.

For the four technologies shown below, f_c is between 10 and 20 Hz for currents on the order of 2–5 mA. We see that it is practically impossible, given the requirements of MIL standards (measured at 1000 hertz), to establish a correlation between irreversible drift and excess noise.

A correlation coefficient $r(t)$ has been calculated,[9] in which t is the duration of the life tests in hours, between the drift $\Delta R/R$ in a group of thin-layer resistors held at 200°C for 40 hours and 1,500 hours, and the excess noise level at a few hertz. Measurements were taken in a population of resistors consisting of approximately 150 to 200 samples. It was estimated that a noise level exceeding three standard deviations from the population of noise recordings indicated abnormal behavior. The resistors were then subjected to life tests for temperature and load. Following these measurements, the correlation between $\Delta R/R$ after 1,500 hours of operation and K measured at $t = 0$ hours, 40 hours, and 200 hours, was established. The results are as follows:

$$r(0) = 0.956 \qquad r(40) = 0.959 \qquad r(1500) = 0.985$$

We see, therefore, that it is indeed possible to estimate the aging of a resistor based on a noise measurement after 40 hours. It is also useful to note that for defective components there is an abnormal noise intensity compared to the population of resistors even at $t = 0$ hours. The value of K is nearly

TABLE 1

Technology	$R(k\Omega)$	$R_L(k\Omega)$	E_0	$I(mA)$	E_1	NI(dB)	$\langle K\rangle 10^{-18}$
Wirewound	10	10	25.75	2	27.30	−50	4.1
Thin-film	10	10	25.75	2	27.80	−49	5.3
Foil	10	10	25.75	2	28.70	−47	7.8
Cermet	10	10	25.75	2	150.0	−33	230

[9]Pinet, D., *op. cit.*

3 orders of magnitude greater than the average for the batch, whereas $\Delta R/R$ after 1,500 hours of operation is only 10 times greater than the average.

We point out that excess noise measurements performed in accordance with the MIL standard at 1000 Hz showed no significant correlation between the values of $\Delta R/R$ and $\langle \delta V^2 \rangle$.

Within a given type of technology, there is also a correlation between the values of K and other physical or geometric parameters. Thick-layer studies have shown that K is inversely proportional to the volume and proportional to the resistivity ρ[10]. The excess noise is also proportional to the volume and, more generally, to the geometric parameters such as the length and width of the resistive element.

To account for these factors, we can express the excess noise spectral density of a resistor T polarized by a current I with the formula:

$$S_e(f) = k_{\text{tech}} \cdot \frac{R_{\blacksquare}}{\text{Vol}} \cdot \frac{I^2R^2}{f} \qquad (V^2/Hz). \tag{1.22}$$

Here, $k_{\text{tech}}(\Omega \cdot L^{-3})$ is a parameter that depends on the technology used to manufacture the resistor R, Vol (L^{-3}) is the volume occupied by the resistive element of the resistor, and $R_{\blacksquare}$ $(\Omega/\blacksquare)$ is the value of the sheet resistance.

3. CONCLUSION

The excess noise in resistive materials reflects the crystal structure of the material and enables us to detect, at least qualitatively, the presence of defects[11] that are responsible for abnormal aging of the component as a function of temperature and time. These measurements are, however, extremely difficult to carry out and quite time-consuming, since they require the use of very low frequencies (a few Hz). At present they cannot be used on production lines but only during quality control operations, using statistical sampling methods.

[10]Whatley, R., "A Model for Low Frequency Noise in Thin Film Resistors." *Proc. Int. Hybrid Micro. Symp.* U.S.A. Vancouver, (1976): 123–127.

Kuo, C. Y. and Blank, H. G. "The Effects of a Resistor's Geometry on Current Noise in Thick Film Resistors." *Proc. 1968 Hybrid Micro. Symp.* Rosmont, P. (1968): 153–160.

[11]These methods are also used in many laboratories to determine aging phenomena by means of the electromigration in conductors. The results are significant enough to enable us to make a correlation between a given mode of aging and a given category of crystal defect. See, for example, Verdier, F. "Caractérisation de l'électromigration: méthodes résistométriques et analyse du bruit en excès." Thesis. University of Bordeaux, France, I, 1993.

6

Heat Transfer

1. INTRODUCTION

In Chap. 1 we saw that a resistance generates heat energy through the Joule effect. In principle this is not "useful"[1] energy in a resistor in which the transfer function consists of supplying an output signal in the form of a potential difference when the input signal is a current and vice versa. There is no internal energy source that can be used to amplify the signal, which is why components of this type are called "passive components." If the resistor constitutes a thermodynamically isolated system, this caloric energy would result in an increase in temperature that would adversely affect its operation.

This energy must, therefore, be evacuated to a heat sink to maintain the resistor's desired characteristics over time.

In this chapter we will study the mechanisms of heat transfer and examine how these concepts can be used to ensure that resistors operate as designed.

[1]Except in the case when the resistive element is used as a source of heat.

2. DESCRIPTION OF HEAT TRANSFER MECHANISMS

There are three different mechanisms by which heat can be transferred from one medium to another: conduction, convection, and radiation. This transfer can only take place from a hot (source) to a cold (sink) medium, as imposed by the first law of thermodynamics. We also know from experiment that the "rate" of heat transfer dQ/dt is proportional to the surface S through which the transfer takes place.

2.1 Conduction

This method of heat exchange tends toward uniform distribution within the medium of the average kinetic energy of the various particles. It occurs through *diffusion and the exchange of kinetic energy arising from collisions among particles composing the medium* and flows from regions in which the average value of this energy—that is, its temperature—is high, toward regions in which it is lower.

The law corresponding to this process of *heat diffusion* states that the density $\vec{\varphi}$ of the heat "current" at a given point, also referred to as the *density of flow,* is a linear function of the temperature gradient at this point:

$$\vec{\varphi}_i = \sum^{j} k_{ij} \cdot \mathrm{grad}_j(T) \qquad (1.1)$$

The operator k_{ij}, which defines the *thermal conductivity* of the medium at the point under consideration, is a symmetric second-order tensor. For an isotropic medium the thermal conductivity (local density of heat flow at a point in the medium) can be simply characterized by a single scalar value k. The vector $\vec{\varphi}_i$ represents the local density of heat flow at a given point in the material medium under consideration. At each point it characterizes the direction and intensity of the density of heat flow.

The sum of all the vectors $\vec{\varphi}$ constitutes a vector field analogous to those encountered in other areas of physics (electric fields, force fields, velocity fields, etc.).

The preceding equation, referred to as *Fourier's law,* can be demonstrated from the laws of statistical mechanics when applied to atoms, molecules, or free electrons in the material under consideration. Conduction is the only method of heat transfer occurring in opaque media.

The coefficient k is expressed as $W \cdot m^{-1} \cdot {}^{\circ}K^{-1}$ in international units when the heat is expressed in joules. Orders of magnitude are between 15 and 420 $W/m \cdot K^\circ$ for metals and metallic alloys, 0.03 and 0.170 $W/m \cdot K^\circ$ for insulators, and 0.007 and 0.170 $W/m \cdot K^\circ$ for gases at atmospheric pressure.

The coefficient k varies with temperature, decreasing for the metals and increasing for gases and insulators. The density of flow $\vec{\varphi}$ represents a power per unit of surface and is thus expressed in W/m^2.

The structure of Fourier's law is similar to that of Ohm's law, potential sources and sinks being replaced by heat sources and sinks. By analogy we use the concepts of tubes and lines of current, isotherms perpendicular to the current lines, and thermal resistances such that $T_1 - T_2 = R_{\text{th}}\phi$ in which ϕ is equal to $\vec{\varphi} \cdot \vec{n}S$, and T_1 and T_2 are isothermal surfaces. R_{th}, therefore, is expressed in $°K/W$. Application of the laws of electricity, using series-parallel combinations for resistance, provides a convenient means for solving many heat problems.

2.2 Convection

Heat transfer obviously occurs regardless of the state of the medium under consideration—solid, liquid, or gas. In the last two cases, however, the significant possibility of deformation in fluids means that different regions of these media have significant freedom of movement at the macroscopic level and can, therefore, transfer an amount of heat directly related to their heat capacity.

We distinguish *forced conduction,* in which the movement of the fluid is produced by an external action (pump, blower, etc.), from *free,* or *natural, convection,* in which movement results from the difference in density between the hot and cold regions of the fluid.

Exact representation of the process of convection transfer presents extremely difficult problems in fluid mechanics, which are not possible to resolve analytically in the majority of cases. However, since we are primarily interested in the quantity of heat exchanged between the fluid and the solid wall around it, we can introduce, by analogy with conduction, a *surface coefficient of heat transfer,* h_c, such that the density of heat flow $\vec{\varphi}$ through a surface element of the wall is proportional to the difference between the temperature T_e of that surface element and the average temperature of the fluid T_f:

$$\vec{\varphi} = h_c \cdot (T_e - T_f) \cdot \vec{n}\,dS \tag{1.2}$$

in which $\vec{n}$ is a unit vector perpendicular to the surface.

The coefficient h_c is expressed in $W \cdot m^{-2} {}^{\circ}K^{-1}$. It varies between 5 and 2,000[2] for the natural convection of air and 280 and 17,000 for the forced convection of water.

2.3 Radiation

Whether spontaneously or through the process of interaction, the atoms, molecules, and free electrons in a body can lose part of their kinetic energy through the emission of electromagnetic radiation (photons). Conversely, when such radiation strikes the surface of a body, a portion is absorbed by the body and is converted to kinetic energy within the medium, in the form of heat.

The fundamental relation is the *Stefan–Boltzmann* equation, according to which the power P of the thermal radiation *emitted* per unit surface of a *black body*[3] is directly proportional to the fourth power of the absolute temperature: $P = \sigma T^4$. σ is the *Stefan–Boltzmann constant.* Its value is $5.6697 \times 10^{-8}\ W \cdot m^{-2} \cdot K^{-4}$.

This expression is derived by integrating over all wavelengths of the fundamental equation obtained by Planck in his quantum theory.

Thermal radiation, like all electromagnetic radiation that does not require any material medium for propagation, represents the only possibility of heat exchange between distant bodies in a vacuum.

When radiating and absorbing surfaces cannot be considered as perfect black bodies, the equation must be modified by a factor F, which is smaller than one, such that:

$$Q\,(\text{watt}) = F\sigma(T_1^4 - T_0^4) \qquad (1.3)$$

When the difference between T_1 and T_0 is small compared with T_1 or T_0, the equation can be linearized:

$$Q = (4F \cdot \sigma \cdot T_1^3)(T_1 - T_0) \qquad (1.4)$$

in such a way that we can define a transfer coefficient:

$$h_r = 4F\sigma T_1^3 \quad \text{and} \quad Q = h_r(T_1 - T_0) \qquad (1.5)^4$$

[2]h_c depends on the temperature and flow conditions of the transfer fluid. Turbulence, for example, strongly increases h_c compared to a laminar flow environment.

[3]We refer to a black body as one that emits the maximum amount of energy through thermal radiation at a given temperature. It is also a body that can absorb all the heat energy it receives.

[4]This equation can only be used if $(T_1 - T_0)/T_0 \ll 1$.

Table 1[5] lists the calculation errors caused by use of the linearized expressions given above in calculating heat transfer by radiation and natural convection (the air temperature is 300°K):

TABLE 1 ($T_0 = 300$°K)

T Source in °K	Natural Convection Air at Standard Pressure $W/°C \cdot m^2$	Radiation $W/°C \cdot m^2$	Error in h_r $W/°C \cdot m^2$	Heat Transfer $W/°C \cdot m^2$	Total Error in %
400	1000	1000	20	2000	1
375	960	660	68	1620	4.2
350	900	390	86	1290	6.6
325	700	170	68	870	8
300	0	0	0	0	0

These results show that, contrary to popular belief, the relative error in heat transfer introduced through the use of the linear approximation of radiative heat transfer is certainly lower than the other causes of measurement error or heat transfer calculation errors introduced by the combination of the materials involved, poorly known contact resistances, and the complex geometric shapes of components.

2.4 Heat Exchange Through Phase Changes

Since any exchange of heat results in a change in the temperature of the bodies involved, it often happens that a phase change (vaporization, condensation, fusion, freezing, etc.) occurs representing a source (or sink) of additional heat. In the ordinary situation where we boil water in a container over a flame, the energy freed by combustion is transferred to the water in the container through:

- Convection of the hot gases of the flame and radiation
- Conduction through the walls of the recipient and fluid layers near the walls
- Convection and conduction within the body of water

When there is sufficient heat, boiling and vaporization occur, these being the essential factors in heat exchange.

[5]Dean, D. J., *Thermal Design of Electronic Circuit Boards and Packages.* Electrochemical Publications Ltd., England, 1985.

Fortunately, we rarely have to determine the contributions from all of the transfer phenomena at the same time, especially in electronic components, where we can generally neglect heat transfer resulting from phase changes. Still, it is important to understand the operating conditions resulting from these various mechanisms.[6]

3. GENERAL EQUATIONS FOR CONDUCTION

Consider a solid medium of volume V. The mechanical and physico-chemical state of this medium is assumed to be invariable. We will assume ρC_p to be its volumetric heat (ρ is the density and C_p the specific heat at constant pressure), k to be its thermal conductivity, and p the power generated per unit volume by internal sources.

3.1 Energy Balance

We apply the first law of thermodynamics to a finite volume v, of surface s contained in a medium V (Fig. 1). Following convention, we consider any energy received by this system as positive.

The quantity of heat exchanged with the exterior per unit of time includes the power exchanged on the boundary surface s and the power

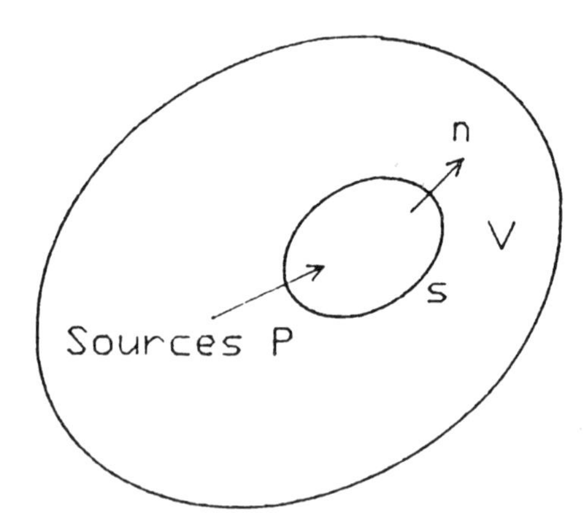

Fig. 1 Finite volume v with surface s contained in a medium V.

[6]Heat pipes are often used to determine transfer rate resulting from phase changes (vaporization, condensation, etc.) in systems electronics.

supplied by the sources. The power crossing s at each point in the direction of the exterior normal $\vec{n}$ is given by $\vec{\varphi} \cdot \vec{n} ds$. Since $\vec{n}$ is the exterior normal at the surface s, the algebraic power received by the volume v crosses S in the direction opposite $\vec{n}$, based on which the power exchanged at the boundary surface s can be expressed as:

$$\int_s - (\vec{\varphi} \cdot \vec{n}) dS \tag{1.6}$$

The power generated by internal sources is given by the integral:

$$\int_V p \cdot dV \tag{1.7}$$

in which p is, as indicated above, the power generated per unit of volume (W/m^3).

Since we assume as a first approximation that the solid cannot be deformed, only local changes of temperature are used in the expression of the instantaneous change in internal energy:

$$\int_V \rho C_p \cdot \frac{\partial T}{\partial t} dV \tag{1.8}$$

The equation representing the first law of thermodynamics can therefore be written:

$$\int_S - (\vec{\varphi} \cdot \vec{n})\, ds + \int_V p dV = \int_V \rho C_p \cdot \frac{\partial T}{\partial t} dV \tag{1.9}$$

Using the Gauss–Ostrogradsky formula, we can transform the surface integral into a volume integral:

$$\int_s - (\vec{\varphi} \cdot \vec{n})\, ds = -\int_V \text{div}\, (\vec{\varphi})\, dV \tag{1.10}$$

The equation representing the first law can now be written:

$$\int_V \left[-\text{div}(\vec{\varphi}) + p - \rho \cdot C_p \frac{\partial T}{\partial t} \right] dV \tag{1.11}$$

which is valid for any point in V.

In a homogeneous and isotropic medium, the thermal characteristics k and ρC_p depend only on temperature. Taking into account Fourier's hypothesis (Eq. 1.1), Eq. 1.11 can be written as:

$$\text{div}\,(k(T) \cdot \overrightarrow{\text{grad}}\, T) + \rho(T) C_p(T) \cdot \frac{\partial T}{\partial t} - p(M,t) = 0 \tag{1.12}$$

in which M is any point within the medium. If we introduce the *thermal diffusivity,* $a = k/\rho C_p$, and the Lapacian Δ, then:

$$\Delta T - \frac{1}{a} \cdot \frac{\partial T}{\partial t} + \frac{1}{k} \cdot \overrightarrow{\text{grad}}\,(kT) \cdot \overrightarrow{\text{grad}}\,T + \frac{p}{k} = 0 \qquad (1.13)$$

To linearize the equation, we have to assume that:

- k and a are independent of T
- p is independent, or a linear function of T, of the form

$$p(M,T,t) = A(M,t) + B(M,t)T$$

For media that are characteristic of resistors, in which p is independent of temperature, we obtain the classic heat equation:

$$\Delta T - \frac{1}{a} \cdot \frac{\partial T}{\partial t} + \frac{p(M,t)}{k} = 0 \qquad (1.14)$$

We can then easily deduce the particular cases that are frequently encountered in practice.

Steady-state environment with internal sources:

$$\Delta T + \frac{p}{k} = 0 \quad \text{Poisson's equation} \qquad (1.15)$$

Steady-state environment without internal sources:

$$\Delta T = 0 \quad \text{Laplace's equation} \qquad (1.16)$$

Nonsteady-state environment without internal sources:

$$\Delta T = \frac{1}{a} \frac{\partial T}{\partial t} \quad \text{Fourier's equation} \qquad (1.17)$$

3.2 Analytical Expressions for the Heat Equation

By introducing the Laplacian expressions in Cartesian, cylindrical, and spherical coordinates, we obtain, respectively:

In Cartesian coordinates:

$$\frac{\partial^2 T}{\partial x^2} + \frac{\partial^2 T}{\partial y^2} + \frac{\partial^2 T}{\partial z^2} + \frac{p(x,y,z,t)}{k} = 0 \qquad (1.18)$$

In cylindrical coordinates, there is the problem of symmetry of revolution with respect to the z axis:

$$\frac{\partial^2 T}{\partial r^2} + \frac{1}{r} \cdot \frac{\partial T}{\partial r} + \frac{\partial^2 T}{\partial z^2} - \frac{1}{a} \cdot \frac{\partial T}{\partial t} + \frac{p(r, z, t)}{k} = 0 \qquad (1.19)$$

In spherical coordinates:

$$\frac{\partial^2 T}{\partial r^2} + \frac{2}{r} \cdot \frac{\partial T}{\partial r} - \frac{1}{a} \cdot \frac{p(r, t)}{k} = 0 \qquad (1.20)$$

4. HEATING IN RESISTORS

In this section we make use of the heat equation to successively examine:

- The manner in which heat is distributed in a cylindrical resistor of radius A and length $2L$ in a steady state and nonsteady state
- The distribution of heat in a flat resistor, referred to as a "chip" resistor, attached by solder or epoxy to an equipotential substrate brought to a temperature T_0

4.1 Heating in a Cylindrical Resistor

A cylindrical resistor is composed of a rod of radius R and total length 2L (Fig. 2). If we assume that, in a steady-state condition, the power in a unit of volume dissipated by the resistor is p (in watt/m^3), then dissipation of this power will take place by convection at the surface and by conduction at the terminals.

The terminals generally have a smaller cross-section than the body of the resistor. Under these conditions dissipation occurs primarily by means of surface convection.

Assume we have a cylinder of radius R (Fig. 2), with thermal conductivity k, placed in a fluid in which surface heat transfer is h_c, the sum of the convection and the linearized radiation (Newton's law), and whose linear dimensions are large compared to its radius. An electric current circulating in this cylinder produces a flow p. Since length $L > R$, we can state that, because of symmetry, the temperature in the cylinder will vary as a function of r alone. Consequently, $\partial T^2/\partial^2 z = 0$. The equation we need to solve will have the form:

$$\frac{1}{r} \frac{\partial}{\partial r}\left(r \frac{\partial T}{\partial r} \right) = \frac{\partial T}{\partial \tau} - \frac{p}{k} \qquad (1.21)$$

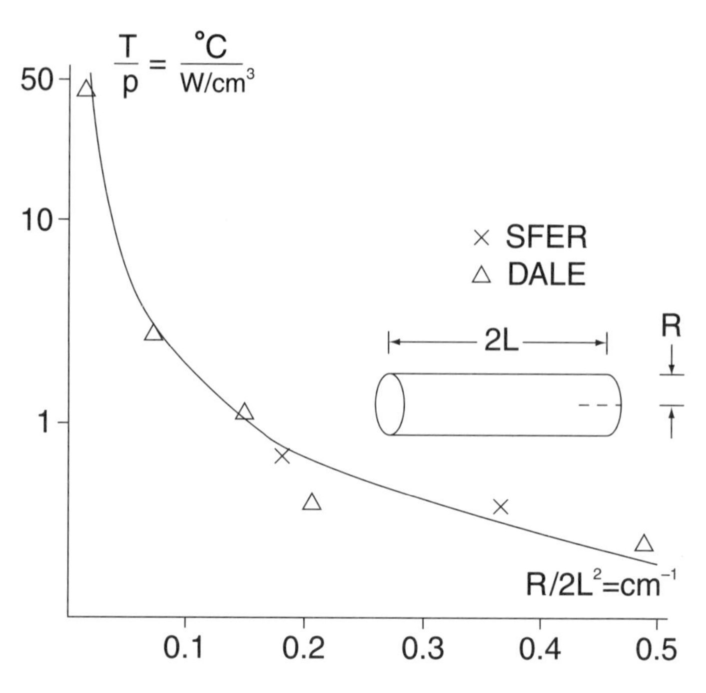

Fig. 2 Thermal resistivity of axial resistors.

in which, to simplify our notation, $\tau = kt/\rho C_p$ (in m^2).

We assume the initial conditions to be $T|_{t=0} = 0$. Boundary conditions at the surface will be determined by the amount of energy supplied by the resistor to the fluid in such a way that heat transfer per unit of surface of the resistor to the fluid can be expressed by:

$$-k \cdot \frac{\partial T}{\partial z}|_{r=R} = h_c T|_{r=R}$$

$$\left(k\frac{\partial T}{\partial r} + h_c \cdot T\right)_{r=R} = k\left(\frac{\partial T}{\partial r} + h \cdot T\right)_{r=R} = 0 \qquad (1.22)$$

Here, $h = h_c/k$, where h_c is the coefficient of dissipation by convection (h, therefore, has the dimension of L^{-1}). By convention we assume that the temperature of the dissipative fluid medium is zero.

Given these boundary conditions, we find the following[7] analytical solution:

$$T = \sum_{n=1}^{\infty} \frac{\vec{T}_n}{\int_0^R r \cdot R_n^2(r) dr} R_n(r) \tag{1.23}$$

in which $\vec{T}_n = \int_0 TR_n(r) \cdot rdr$ and $R_n(r)$ is an eigenfunction of the equation that satisfies the following differential equation:

$$\frac{1}{r}(rR')' + \lambda R = 0 \tag{1.24}$$

with the boundary conditions being:

$$R'(R) + h \cdot R(R) = 0$$

and where $R(0)$ is finite. A solution for these conditions exists if and only if:

$$\lambda = \lambda_n = \frac{\gamma_n^2}{R^2} \quad n = 1, 2, \ldots$$

in which the values γ_n are the consecutive positive roots of the equation $\gamma J_1(\gamma) = AhJ_0(\gamma)$.

The solutions satisfying Eq. 1.24 are:

$$R = R_n(r) = J_0\left(\frac{\gamma_n r}{R}\right)$$

These functions are orthogonal in the interval $(0,R)$ and:

$$\int_0^R rR_n^2(r)\, dr = \frac{R^2}{2}[J_0^2(\gamma_n) + J_n^2(\gamma_n)] = \frac{R}{2} J_n^2(\gamma_n)\left[1 + \left(\frac{hR}{\gamma_n}\right)^2\right] \tag{1.25}$$

The function T_n is obtained by multiplying Eq. 1.21 by R_n and integrating from 0 to R. We then integrate by parts twice and, by taking into account the boundary conditions, we obtain:

$$T_n' + \left(\frac{\gamma_n}{R}\right)^2 \cdot T_n = \frac{pR^2}{k} \frac{J_1(\gamma_n)}{\gamma_n} \tag{1.26}$$

[7]Lebedev, N. N. *et al., Problems of Mathematical Physics.* p. 121, ex. 247. Englewood Cliffs, N.J.: Prentice-Hall, 1965.

Martinet, J., *Thermocinétique Approfondie.* Paris: Tec.Doc. Lavoisier, 1990.

The solution of Eq. 1.26 that satisfies the condition $T_n = 0$ is:

$$\vec{T}_n = \frac{pR^4}{k} \frac{J_1(\gamma_n)}{\gamma_n^3} \{1 - e^{-\gamma_n^2 \tau / R^2}\} \quad (1.27)$$

By introducing Eqs. 1.25 and 1.27 in Eq. 1.23, we obtain the following solution:

$$T(r, \tau) = \frac{2pR^2}{k} \sum_{n=1}^{\infty} \frac{J_1(\gamma_n) \cdot J_0(\gamma_n r/R)}{\gamma_n^3 \cdot [J_0^2(\gamma_n) + J_1^2(\gamma_n)]} \cdot [1 - e^{-\gamma_n^2 \tau / R^2}] \quad (1.28a)$$

This solution can be applied for low values of τ; that is, when the current in the resistor is first established. For high values of τ it is preferable to eliminate the terms that are independent of time, by using the formula:

$$\frac{1}{8}\left(1 - \frac{r^2}{R^2}\right) + \frac{1}{4Rh} = \sum_{n=1}^{\infty} \frac{J_1(\gamma_n) \cdot J_0(\gamma_n r/R)}{\gamma_n^3 [J_0^2(\gamma_n) + J_1^2(\gamma_n)]}$$

in which the first term is the solution of the differential Eq. 1.21, independent of time ($t \Rightarrow \infty$). Under these conditions, $T(r, \tau)$ assumes the form of Eq. 1.28b.

$$T(r, t) = \frac{2pR^2}{k} \left[\frac{1}{8}\left(1 - \frac{r^2}{R^2}\right) + \frac{1}{4Rh} - Rh \sum_{n=1}^{\infty} \frac{e^{-\gamma_n^2 \tau / R^2} \cdot J_0(\gamma_n r/R)}{\gamma_n^4 [1 + (hR/\gamma_n)^2] \cdot J_0(\gamma_n)}\right] \quad (1.28b)$$

As noted above, Eq. 1.28b gives the temperature of a cylinder of infinite length. It would, therefore, be useful to calculate the conditions for establishing temperature as a function of time at the hottest point, that is, for $z = 0$. When the cylinder has a length of $-L \leq z \leq L$, the additional boundary conditions at the ends provide the solution for a steady-state environment:[8]

$$T(r, z) =$$

$$\frac{2phR^3}{k} \sum_{n=1}^{\infty} \frac{1 - \dfrac{(hR/\gamma_n)\cosh(\gamma_n z/R)}{[\tanh(\gamma_n L/R) + (hR/\gamma_n)]\cosh(\gamma_n L/R)}}{[1 + (hR/\gamma_n)^2]\gamma_n^4 J_0(\gamma_n)} \cdot J_0\left(\frac{\gamma_n r}{R}\right) \quad (1.29)$$

in which the values of γ_n are the positive roots of the equation:

$$\gamma J_1(\gamma) = Rh J_0(\gamma)$$

[8]Lebedev *et al. op. cit.*, p. 129, ex. 268.

Given the limited dimensions of actual components, we can neglect the terms of the series in Eq. 1.29 for values of $n > 2$ or 3, depending on the circumstance.

For example, consider a ceramic power resistor with $R = 0.1$ cm and half-length $L = 0.3$ cm.

The resistor dissipates through natural convection and radiation in such a way that $h_c = 50\ W \cdot m^{-2} \cdot K^{-1}$ (air at rest at 25°C). Then h is equal to $50\ m^{-1}$. Thermal conductivity of the ceramic is $k = 1\ W \cdot m^{-1} \cdot K^{-1}$.

Under these conditions, $\gamma_1 \cong 0.48$, $\gamma_2 \cong 4$, $\gamma_3 \cong 7$, and $\gamma_4 \cong 10$, and the temperature at the surface is represented by:

$$T(r,z) = 100pR^3 \cdot \frac{1 - \dfrac{R\cosh(0.47\ z/R)}{\tanh(0.47\ L/R + R\cosh(0.47\ L/R)}}{(1 + R_2) \cdot 0.048 J_0(0.47)} \cdot J_0(0.48\ r/R) \tag{1.30}$$

Calculations show that[9] the maximum temperature at $z = 0$ is 100°C, the temperature at $z = 0.3$ cm being approximately 24°C. We then can expect from this resistor a $p = 63 W \cdot \mathrm{cm}^{-3}$ and a nominal power $\mathrm{Pw} = p \cdot 2\pi R^2 L \cong 1.25$ w. The dimensions of this resistor are approximately those of the model DALE RWR81 (nominal power 1W). For model RWR80, Eq. 1.30 leads also to a power of 2.3W (nominal power 2W).

Through Eqs. 1.28 and 1.29, we can calculate a thermal resistivity curve based on that shown in Fig. 2. The experimental points are results of dissipation measurements performed by DALE Corp. on resistors type RWR. We see that correlations with mathematical model is quite reasonable.

The conductivity of the terminal wires can be neglected in this example. Assume that the wires are copper, with a length $1 = 3$ cm and a cross-section $s = 3.14$ mm². For the two wires, Fourier's equation gives $(\Delta T = 24°\mathrm{C})$:

$$Q(\mathrm{W}) = Sk\Delta T/L = 3.14 \cdot 10^{-6} \times 392 \times 24/3 \cdot 10^{-2}$$
$$= 1\mathrm{W} \text{ for each wire.}$$

Energy dissipation in cylindrical resistors with output terminal wires occurs primarily through convection and radiation.

[9]Cherbuy, J., Personal communication.

4.2 Heating in Flat Resistors

Resistors in the form of rectangular chips are mainly used in electronic systems because of their similarity in shape to active components and the frequent use of *surface-mount component technology.*

Two technologies are used to mount resistors on printed circuits:

1. A flat component covered with a protective material and supplied with two (or more) output wires, mounted as shown in Fig. 3
2. A flat resistive component, supplied with two or more terminal pads, to which gold or aluminum wires are attached by means of thermal bonding, as shown in Figure 4[10]

4.3 Components as Shown in Fig. 3

A significant part of the energy will be lost by convection (and radiation) on the surfaces of the resistor. Calculation of excess temperature is relatively straightforward. Assume a thermal wall (Fig. 5) of thickness[11] $2e$ and width $L >> e$, in which a source supplies p (W/m^3). The problem can

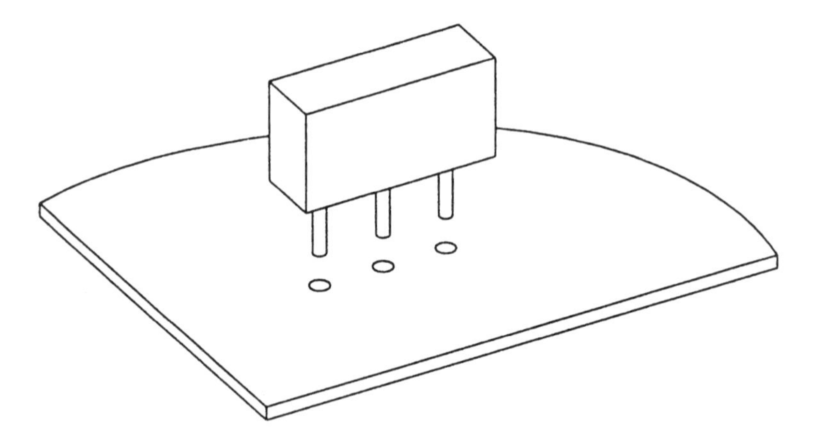

Fig. 3 Single in-line resistor array mounted on a PCB.

[10]There are several other methods of connecting components in addition to that shown in Fig. 4. These include the use of contact pads located on the lower surface of the substrate (flip chip) or on the perimeter of the component (chip carrier).

[11]This condition can be used to reduce the complex three-dimensional problem to a problem in one dimension by assuming that the component behaves like a semi-infinite wall.

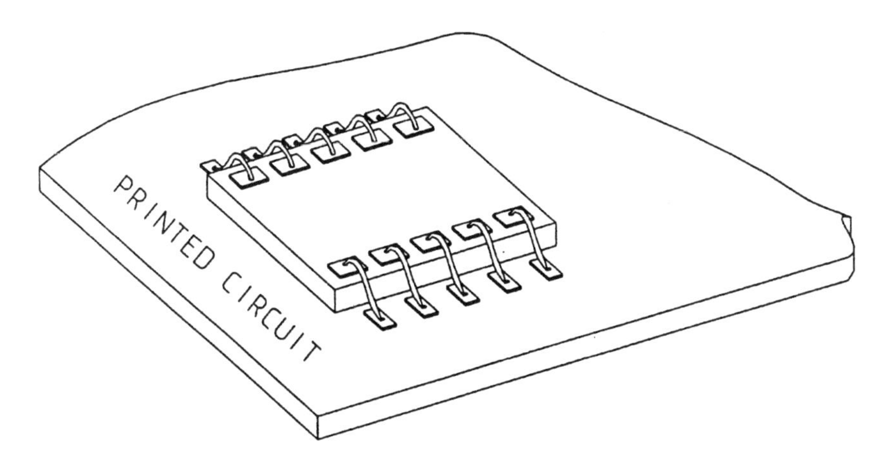

Fig. 4 Chip resistor interconnected by wire bonding.

be reduced to integrating the equation $\partial^2 T/\partial x^2 = \partial T/\partial \tau$ in which, as in the previous case, $\tau = kt/\rho C_p$.

Given the initial condition of $T|_{\tau=0} = 0$ and the boundary conditions

$$\left| k \frac{\partial T}{\partial x} + h_c T \right|_{x=-e \text{ and } x=+e} = 0 \tag{1.31}$$

along with the integral formula for the half-thickness, we obtain:

$$\frac{d}{dt} \int_0^e \rho C_p \cdot T(x,t) dx - k \frac{\partial T}{\partial x}(e,t) = pe \tag{1.32}$$

Since in a nonsteady-state environment the distribution of temperature assumes the shape of the distribution in a steady-state $\partial T/\partial t = 0$ it is

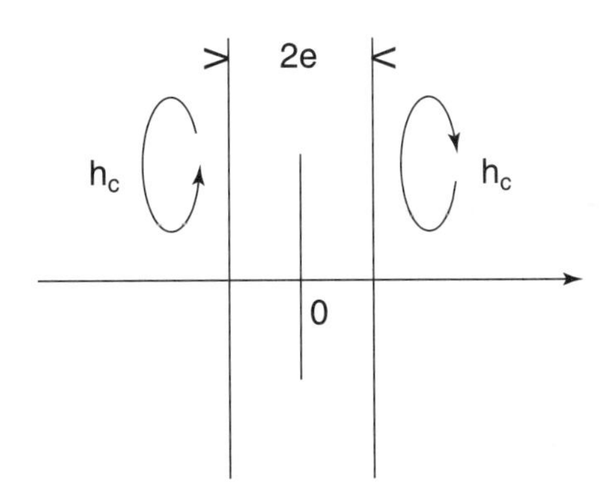

Fig. 5 Schematic of a thermal wall.

easy to show,[12] by assigning $h=h_c/k$ as we did above, that the equation satisfies a solution of the type:

$$T(x, \tau) = \frac{pe^2}{2k}(1 - x^2/e^2 + 1/2he) \cdot u(\tau)$$

in which the expression in parentheses is the steady-state solution and $u(\tau)$ is a function to be determined, given that $u(0) = 0$. By placing the above expression in the integral formula, we obtain:

$$\frac{du(t)}{dt} + \frac{k}{e^2\rho C_p} \cdot \frac{3he}{he + 3} \cdot (u(t) - 1) = 0$$

and, after solving for $u(t)$:

$$T(x,t) = \frac{pe^2}{2k}\left(1 - \frac{x^2}{e^2} + \frac{2}{eh}\right) \cdot \left(1 - \exp\left(-\frac{3eh}{eh+3} \cdot \frac{\tau}{e^2}\right)\right) \quad (1.33)$$

In a steady-state environment the maximum temperature T_{max} will be, as anticipated, at the abscissa $x = 0$ and $\tau = 0$:

$$T_{max} = \frac{pe^2}{2k} \cdot \left(1 + \frac{2}{eh}\right) \quad (1.34)$$

These equations enable us to determine the thermal resistance from convection and radiation R_{thc} (in $°K \cdot W^{-1}$) when the active volume V of the component is known:

$$1/R_{thc} = \frac{2kV}{e^2} \cdot \left(1 + \frac{2}{eh}\right)^{-1} \quad (1.35)$$

In this component, having n leads of cross-section S, length l, and thermal conductivity k_s, the heat resistance R_{ths} of the leads will be equal to:

$$R_{ths} = \frac{1}{n \cdot k_s \cdot S} \quad (1.36)$$

Equations 1.33–1.36 enable us to calculate the overall thermal behavior of a flat resistor mounted perpendicularly to the printed circuit or substrate plane.

In effect the total heat resistance R_{thT} of the component can be expressed by:

$$R_{thT} = \frac{R_{thc} \cdot R_{thS}}{R_{thc} + R_{thS}} \quad (1.37)$$

[12]Lebedev, N. N., *et al. op. cit.*

Let us assume that we have a Vishay type foil S102 resistor of dimensions[13] $e = 1$ mm and sides 5 mm, standing in air such that $h_c + h_r = 45$ $W \cdot m^{-2} \cdot K^{-1}$, and $k = 0.8$ $W \cdot m^{-1} \cdot K^{-1}$. The active part is connected to the leads (copper, of cross-section 1 mm2 and length 5 mm) by two Cupron or manganin ribbons ($k \cong 200$ $W \cdot m^{-1} K^{-1}$) of thickness 25 μm, width 0.3 mm, and length 5 mm. Under these conditions the total heat resistance R_{thT} is 385 °C/W with $R_{thc} = 500$ °C/W and $R_{ths} = 1672$ °C/W.

If the maximum allowable operational temperature at nominal power is 125°C, then the resistor can dissipate approximately 0.3 watts of power, which is the value indicated by the manufacturer.

During the fabrication cycle for the resistor, it is necessary to overload it with 5 times the nominal power, or 1.5W, for a period of 5 seconds. Under such dynamic conditions, the coefficient h varies rapidly, reaching a value of 150 after a few milliseconds. The maximum temperature rises to approximately 170°C above room temperature.

Using the above equations we can calculate, *as a first approximation*, the order of magnitude of the maximum temperatures in flat components whenever the thickness e is considered small compared to the other dimensions.

4.4 Components as Shown in Fig. 4 (see page 99)

These chip resistors are used in hybrid microelectronic circuits, in which active chips and passive, resistive, capacitive, or inductive components are glued with epoxy resins to ceramic substrates or printed circuits, and then connected to provide complex and highly specialized electronic functions.

Given the economic and technological interest in this technology, numerous books and articles have been published[14] as part of the technical literature for calculating the heat dissipation from sources on a thin substrate containing a heat sink on its rear surface.

[13]The dimensions of the component are such that the equilibrium temperature is reached immediately and we can neglect the terms in the sum of Eq. 1.33.

[14]The following are among the "basic" documents in the specialized literature on the subject:

Kennedy, D. P. J., *Applied Physics.* 31, 8 (1960): 1490.

Linstead, R. D., *et al., IEEE Trans. on Elect. Dev.* 19(1) (1973): 51.

Dean, D., *Handbook of Thick Film Technology.* Chapter 11. Electrochemical Publishers.

Dean, D., *Thermal Design of Electronic Packages.* (*op. cit.*).

David, R. F., *Proc. 27th Electron. Conf.* (1977): 324.

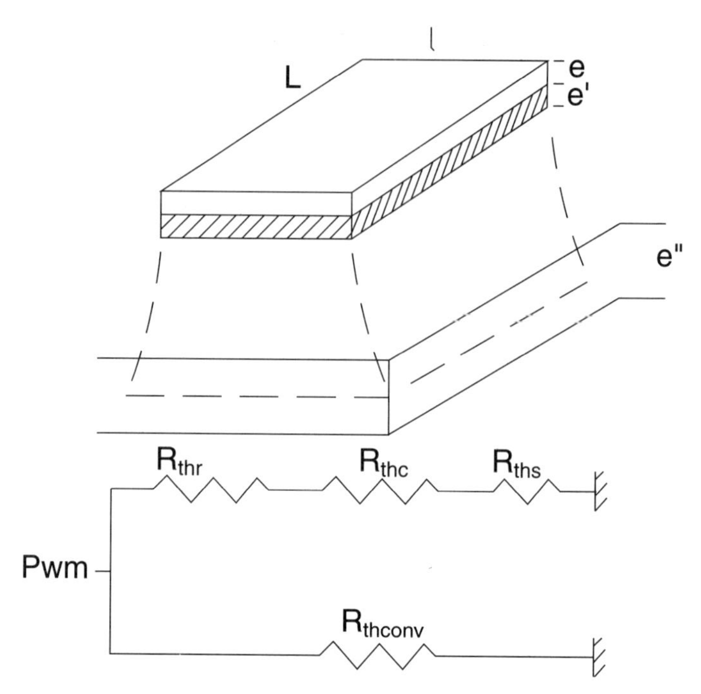

Fig. 6 Thermal resistance of a chip glued to a substrate.

Figure 6 shows a flat resistor glued to a substrate by means of an epoxy. The thickness of the resistive chip is e and its thermal conductivity is k_r. The thickness of the epoxy is e' and its thermal conductivity is k_c. The thickness of the substrate is e'' and its thermal conductivity is k_s. The heat sink is maintained at a temperature $T_\infty = 0$. The constant h_{conv} is $50\ W \cdot m^{-2} \cdot K^{-1}$. The length and width of the chip are L and 1. Using this information, we can calculate the heat resistances:

- R_{thr} of the chip is $e/(k_r L \cdot 1)$
- R_{thc} of the epoxy is $e'/(k_c L \cdot 1)$
- Thermal R_{ths} of the substrate is $e''/K_s(L + e'')(1 + e'')$

The optimum conditions of heat resistance were determined by D. J. Dean.[15] When the thickness of the substrate is the same magnitude as the

[15]Dean, D. J. *op. cit.* p. 242, supplement S.5.

chip dimensions, as in the case of chips bonded to a printed circuit board, the lateral diffusion of heat in the chip must be taken into account. Dean calculated that:

$$R_{\text{ths}} = \frac{1}{2k_s(l - L)} \cdot Ln\left(\frac{l(2e'' + L)}{L(2e'' + l)}\right) \tag{1.38}$$

and, for a square chip, $l = L$, $R_{\text{ths}} = e''/[k_s(L + 2e'')L]$

- Thermal resistance by convection:

$$R_{\text{thconv}} = 1/(\beta \cdot h_c L \cdot l) \tag{1.39}$$

in which $\beta = 1.06$ to take into account convection across the surface of the component. Let us call R_{thcomp} the sum:

$$R_{\text{thcomp}} = R_{\text{thr}} + R_{\text{thc}} + R_{\text{ths}} \tag{1.40}$$

The total heat resistance R_{THTOT} of the chip unit plus substrate will be:

$$\frac{1}{R_{\text{THTOT}}} = \frac{1}{R_{\text{thconv}}} + \frac{1}{R_{\text{thcomp}}} \tag{1.41}$$

Application

$L = 4$ mm, $l = 3$ mm, e $= 0.3$ mm, $k_r = 25\ W \cdot m^{-1} \cdot °K^{-1}$ (ceramic)

$e' = 25\ \mu$m, $k_c = 0.2\ W \cdot m^{-1} \cdot °K^{-1}$ (pure epoxy adhesive)

$e'' = 2.5$ mm, $k_s = 2\ W \cdot m^{-1} \cdot °K^{-1}$ (printed circuit)

$h_c = 30\ W \cdot m^{-2} \cdot °K^{-1}$

Under these conditions, $R_{\text{THTOT}} \approx 50$ °C/W. If we assume a maximum temperature of 100°C above room temperature, the chip will dissipate 2W.

4.5 Transfer at the Interfaces and Contact Resistances

The examples above indicate the methods of mathematical analysis that enable us to calculate dissipation in components (active or passive). However, experiments show that the results obtained are realistic only in the simplest cases, such as those involving cylindrical resistors, and that as soon as the component consists of an assembly of various materials (ceramic, metal, epoxy, etc.) separated by interfaces, it is the nature and quality of the interfaces that determine heat transfer.

In practice the heterogeneous nature of the materials used introduces structural defects (voids, oxidation, etc.) that constrict heat flow and sometimes strongly increase heat resistance. Figure 7 (a and b) illustrates this phenomenon.[16]

Let us assume a cylindrical tube of cross-section S, whose walls are insulated except for a surface s, to which we apply power W. We can show that, if the base is at temperature T_B, the average temperature $\langle T_0 \rangle$ along s is given by the equation:

$$\langle T_0 \rangle = T_B + \left\{ \frac{1}{k}\frac{L}{S} + \frac{1}{k} g \right\} W \tag{1.42}$$

in which $r_{\text{th}} = (1/k)g$ represents the increase of resistance in the medium due to the constriction of the lines of flux. g is the constriction and depends only on the geometry (shape of the tube, s, and S).

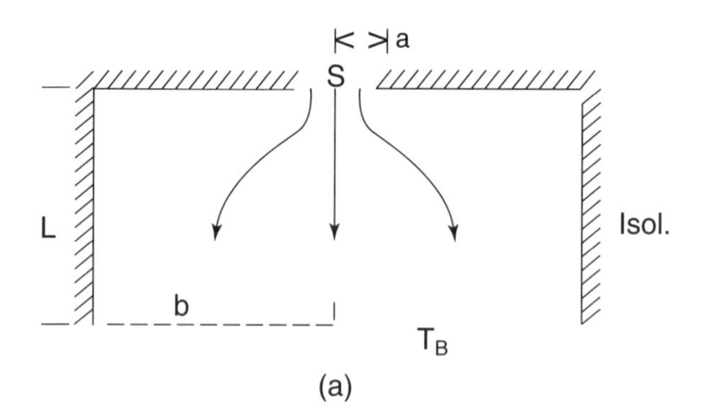

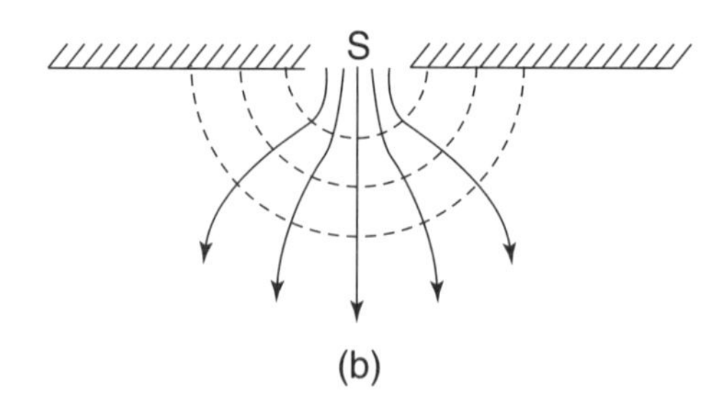

Fig. 7 Constriction of the heat flow.

[16]Bardon, J. P. and Scudeller, Y., "Champs thermiques et phénomènes de couplage au sein des circuits hybrides de puissance." *Journées I.S.H.M.* Bordeaux, October 28, 1992.

If s and S are circular and have radii of a and b, we then have:

$$g = \frac{8}{3\pi^2 a}\left\{1 - 1.305\frac{a}{b} + 0.246\frac{a^3}{b^3}\cdots\right\} \tag{1.43}$$

When the medium is semi-infinite (Fig. 7b), then $b \Rightarrow \infty$, $T_B \Rightarrow T_\infty$, $g \Rightarrow f = 8/3\pi^2 a$ and $T_0 = T_\infty + (1/k) \cdot fW$; $f \Rightarrow g$ is the constriction function in an infinite medium. It can be shown that, for constriction in a semi-infinite medium, 96% of $T_0 - T_\infty$ is confined within a hemisphere of radius $10a$.

Calculations and experimental[17] results show that we can define critical values for thermal resistance at an interface. The orders of magnitude are given in Table 2.

To illustrate this, we will consider a resistive component as shown in Fig. 6, and assume that its construction conforms to that shown in Fig. 8. When we take into account the heat resistance of the interfaces by using the average values given in Table 2, the resistance R_{thcomp} increases from 100 K/W to approximately 160 K/W. This means the total heat resistance R_{THTOT} is approximately equal to 57°C/W, a difference of 14% for a component with only three interfaces. The heat resistance is thus highly dependent on the number of interfaces, but also depends on their characteristics and, consequently, on the growing complexity of the component. We will see in the chapter on power resistors (Chap. 9) that dissipation in a surface-mounted power component can double depending on the interface characteristics.

In reality, since the total heat sink supplied by the printed circuit is not infinite, we find that we are soon limited by other sources of heat from

TABLE 2

Type and Condition of Contact	Contact Resistance $(K \cdot W^{-1} \cdot m^2)$
1 – *Pressed media*	
Roughness and rippling	10^{-4} to 10^{-3}
Ripple-free surfaces	10^{-5} to $2 \cdot 10^{-5}$
2 – *Glued or brazed media*	
Impurity deposits	10^{-6} to $5 \cdot 10^{-5}$
Adhesion defects	10^{-6} to $5 \cdot 10^{-5}$
3 – *Close contact media*	
Metal deposits	$\approx 10^{-7}$
Solder	$\approx 5 \cdot 10^{-7}$

[17]Scudeller, Y., I.S.I.T.E.M. Personal communication.

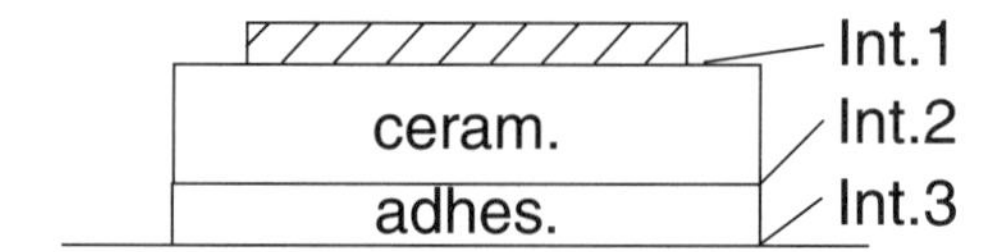

Fig. 8 Construction of a resistive component per Fig. 6.

neighboring chips. Under these conditions the resistance R_{ths} is much greater. If chips with an average dimension of 4 x 3 mm^2 are separated by approximately 3 mm, then we can estimate that the heat resistance of the printed circuit, without other means than natural convection, will be approximately 1500 °C/W(!) and the power dissipated per mounted chip must be limited to 0.1W.

5. CONCLUSION

In this chapter we have shown that a first-order estimate of heat transfer in a resistor is relatively simple in all cases. More accurate calculations require the use of finite element numerical methods. Nevertheless, we have seen that estimates of the heat dissipation of a component during operation depend on several external factors. Among the most important are the coefficient of exchange by convection. These factors depend on the nature of the fluid involved, the type of convection (natural or forced), etc. The materials used for heat sinks will also have a considerable effect on the results.

The user must therefore always pay careful attention to the environment in which these components operate. Otherwise, considerable problems may arise if components are used, even at their rated power, in an unsuitable environment. Calculations must also be checked by measurements of heat dissipation during operation. We will describe such calculations in Chapter 9, which covers power resistors.

PART II

Technology

INTRODUCTION

In Part I we described in detail the physical principles relating the dependence of performance and the electrical and mechanical characteristics of resistive components to the materials and fabrication processes used in their production. Part II concerns the purely technological and practical aspects of these components. Using documentation provided by Vishay Intertechnology, Inc. and other manufacturers, we will provide an exhaustive examination of the criteria for selecting resistive components for use in specific applications.

While the information found in Part II is based on the theoretical foundation provided in Part I, this section of the book can be used independently. Those readers who would like a fuller understanding of the theoretical principles governing the design of a given component may wish to consult the corresponding chapter in Part I.

Chapter 7 introduces the terminology commonly used to describe the characteristics and performance of resistive components, fixed and variable (such as potentiometers and trimmers).

The chapters following describe in greater detail the manufacturing methods, materials used, and the expected performance of the above-mentioned and other (nonlinear and magnetoresistive) components. Using the theoretical developments of Part I as a base, the technological choices (materials, geometrical forms, processes, etc.) are examined in detail for optimization of performance and reliability.

A number of appendices were added dealing with special subjects such as film resistors with bulk properties (paper that fathered the foil technology), methods for quality assurance, models for accelerated lifetime testing, and resistive strain gages.

7

Overview of Fixed and Variable Resistors

1. FUNDAMENTAL PARAMETERS

1.1 Ohmic Value

Resistors are passive components (unlike transistors, integrated circuits, etc.), characterized by the deposition of a resistive element on an appropriate substrate. This resistive element can be a wire, metallic layer, metal foil, etc. Its ohmic value is determined by means of the following formula:

$$R = \rho L/S$$

where $L(\mathrm{m})$ = length of deposit
$R(\Omega)$ = resistance
$\rho(\Omega\mathrm{m})$ = resistivity of the material forming the deposit
$S(\mathrm{m}^2)$ = cross-section of the deposit

R is inversely proportional to thc thickness of the resistive material, which, for obvious technical reasons, must be greater than zero (the diameter of wound wire is close to 10 μm; the thickness of a metallic deposition layer is roughly a micrometer). This relationship becomes more significant as the desired power increases, limiting the range of ohmic values possible with these devices.

1.2 Nominal Ohmic Value of a Resistor

Nominal value is the desired ohmic value of the resistor. Standard preferred numbers of nominal values R_n are obtained by rounding the theoretical value obtained:

$$R = \sqrt[m]{10^n}$$

where n is a positive or negative integer number

m can take the values 3, 6, 12, 24, 48, 96, or 192

The series obtained (E3 to E192) include between 3 and 192 values per decade. They are associated with increasingly closer tolerances (greater than ±20% for the E3 series, ±0.5% for the E192 series).

Table 1 lists, for $n = 1$, the nominal values and tolerances for the series E6, E12, and E24.

TABLE 1 Nominal Current Values and Related Tolerances

Series E24 Tolerance ±5%	Series E12 Tolerance ±10%	Series E6 Tolerance ±20%
1.0	1.0	1.0
1.1		
1.2	1.2	
1.3		
1.5	1.5	1.5
1.6		
1.8	1.8	
2.0		
2.2	2.2	2.2
2.4		
2.7	2.7	
3.0		
3.3	3.3	3.3
3.6		
3.9	3.9	
4.3		
4.7	4.7	4.7
5.1		
5.6	5.6	
6.2		
6.8	6.8	6.8
7.5		
8.2	8.2	
9.1		

TABLE 2 Marking Codes and Resistance Values

Resistance	Code
0.1 Ω	R10
0.332 Ω	R332
0.590 Ω	R59
1 Ω	1R0
3.32 Ω	3R32
10 Ω	10R
33.2 Ω	33R2
100 Ω	100R
1 kΩ	1K0
10 kΩ	10K
100 kΩ	100K
1 MΩ	1MO

When the required tolerance is small (±0.005% for foil resistors, for example), it is important to follow the terms of the E192 series as closely as possible. There are times, however, when the selection of nominal resistance R_n is, by necessity, completely random.

R_n is the value indicated on the body of the resistor or potentiometer; the value can be represented directly or by means of a numerical (Table 2) or color (Table 3, see next page) code.

1.3 Dissipation and Nominal Temperature

The maximum power that can be continuously applied to a resistor depends on the ambient operating temperature (it is important to consider the temperature change in the component during operation, which increases the ambient temperature).

Each resistor model is associated with a curve, known as a derating curve, which indicates the maximum allowable dissipation as the ambient temperature changes.

To simplify their use, these curves are generally shown as two straight-line segments, temperature and nominal dissipation corresponding to point A on the curve shown in Fig. 1.

The International Electrotechnical Commission (IEC) recommends that the nominal temperature be determined for all models at 70°C, and that

TABLE 3 Color Coding

Color	Significant Figures	Temperature Coefficient	Tolerance	Temperature Coefficient (10^{-6}/°C)
Silver	—	10^{-2}	±10%	—
Gold	—	10^{-1}	±5%	—
Black	0	1	—	±250
Brown	1	10	±1%	±100
Red	2	10^{2}	±2%	±50
Orange	3	10^{3}	—	±15
Yellow	4	10^{4}	—	±25
Green	5	10^{5}	±0.5%	±20
Blue	6	10^{6}	±0.25%	±10
Violet	7	10^{7}	±0.1%	±5
Gray	8	10^{8}	—	±1
White	9	10^{9}	—	—
None	—	—	±20%	—

nominal power dissipation be defined as the maximum power that can be continuously applied at an ambient temperature of 70°C.

With rare exception the stated nominal temperature is either 25, 70, or 125°C. It should be pointed out that the minimum temperature for this category of component is the minimum ambient temperature at which the

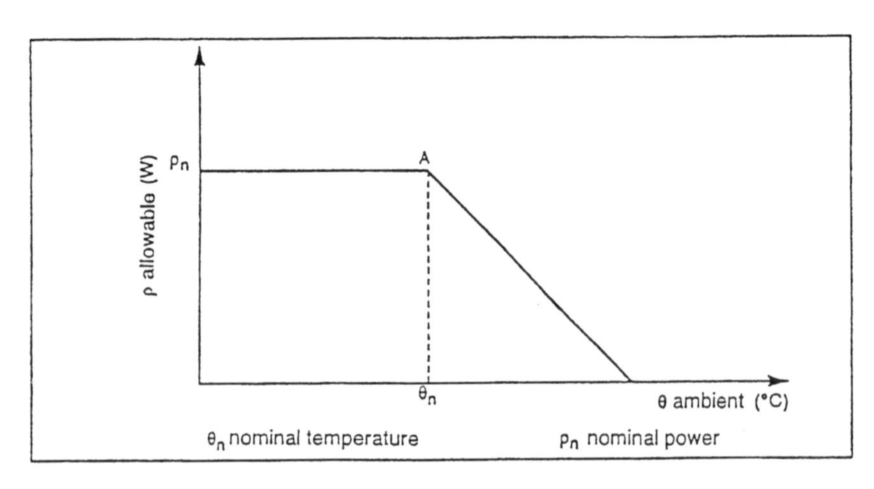

Fig. 1 Allowable dissipation as a function of ambient temperature θ.

resistor is designed to operate continuously at a fraction of its nominal dissipation, which can be zero.

1.4 Critical Resistance

A given resistor with nominal power Pn and maximum allowable voltage Umax has a single resistance value at which it dissipates nominal power at the maximum voltage. This value is generally referred to as the *critical resistance.*

For resistance values lower than the critical resistance, the maximum voltage is never reached, while for values greater than the critical resistance, the applied power diminishes with nominal power. Figure 2 illustrates this relationship.

1.5 Temperature Coefficient

The ohmic value of a resistor R_0 is determined at a reference temperature θ_0 of 20° ± 5°C. The ohmic value of the resistor varies with the temperature of the resistive element. Figure 3 illustrates this variation. The slope p of the straight line I (chord slope) represents the value of the temperature coefficient TCR_θ of this resistor at temperature T.

$$\mathrm{TCR}_\theta = \frac{R_\theta - R_0}{R_0} \cdot \frac{1}{T - T_0}$$

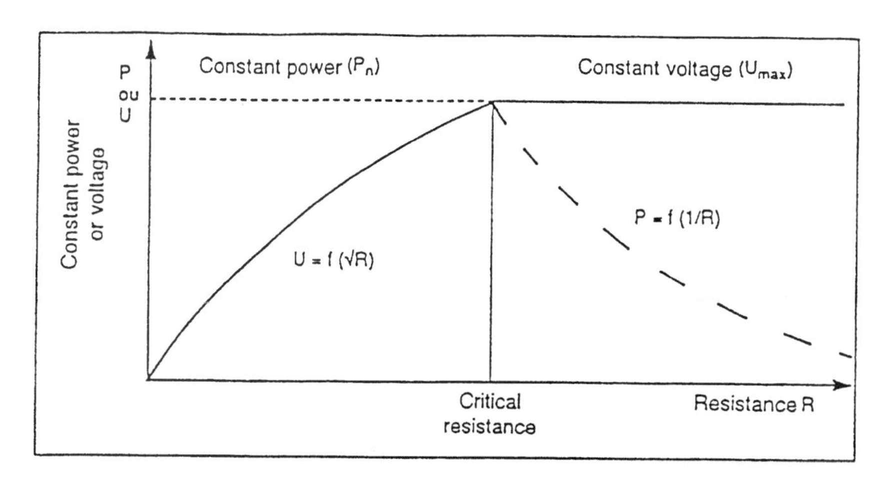

Fig. 2 Critical resistance.

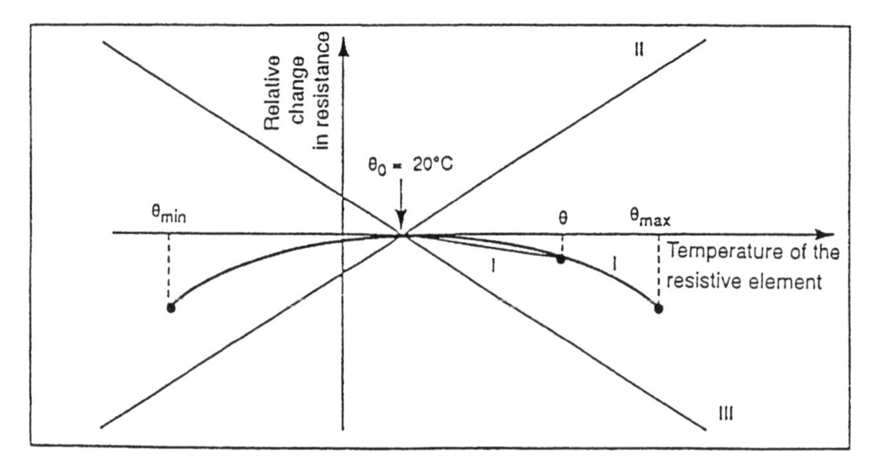

Fig. 3 Temperature coefficient.

where R_θ = ohmic value of the resistor at temperature θ
R_0 = ohmic value of the resistor at temperature θ_0

The temperature coefficient TCR_θ is generally expressed in parts per million per degree Celsius. For a given resistance the maximum value of the temperature coefficient represents the maximum value of the slope of the line I when the temperature T varies between the two temperature limits defined for this resistor model. For each resistor model there are two guaranteed temperature limits. If we draw two straight lines, II and III, on Fig. 3, whose slopes are equal at these limits, all curves T associated with resistors of the same model will fall within the region determined by II and III.

For some resistor models, however, the limits of the guaranteed temperature coefficient depend on the resistor's nominal ohmic value.

1.6 Voltage Coefficient

The voltage coefficient of a resistor is defined as the variation in ohmic value as a function of the applied voltage. Its influence is generally negligible for the majority of technologies because of its low value ($< 10^{-6}$/V) and the limitation that it be maintained to within a few hundred volts of the applicable voltages. Its effect is really significant only in resistors designed for use in situations where voltages can reach several thousand volts or in very high precision resistors.

2. RESISTORS

2.1 General

Depending on the technology, we can classify resistors according to several parameters, which depend on the specific needs of users. The most common are ohmic value, tolerance, stability during storage or operation (at a desired precision), and dissipation.

Table 4 provides an overview of several resistor technologies as a function of the required precision.

2.2 Wirewound Resistors

2.2.1 Technology

Resistors are most frequently made by winding a metal wire around a ceramic, plastic, or fiberglass core. The ends of the wire are soldered to two caps, which are inserted onto the ends of the core. The assembly is protected

TABLE 4 Breakdown of Resistor Types as a Function of Precision

Class	Resistive Element	Tolerance %	Stability %	Maximum Temperature Coefficient (10^{-6}°C)
High precision	Wirewound Metal foils Thin film	0.005, 0.01, 0.025, 0.05, 0.1	0.05, 0.1	±5, ±10
Precision	Wirewound Metal film	0.1, 0.25, 0.5	0.25, 0.5	±10, ±25
Semiprecision	Wirewound Metal film Metal oxide film Pyrolytic carbon film Cermet	1, 2, 5	1, 2, 5	±50, ±100, ±100{ +200 / −500
Standard	Wirewound Pyrolytic carbon film	5, 10, 20	5, 10, 20	+200, ±1000 { −500

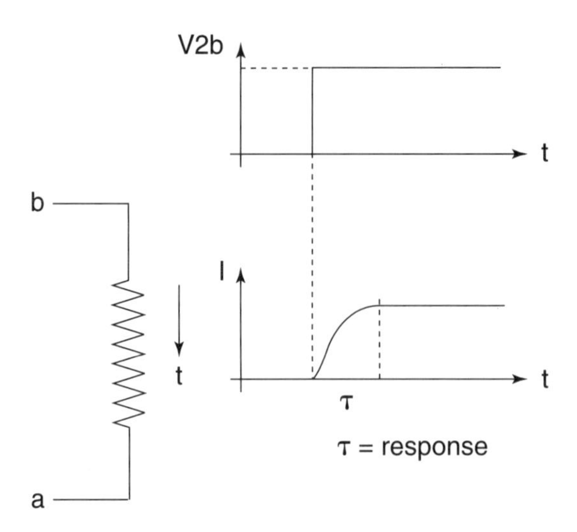

Fig. 4 Time response to a voltage pulse.

with a layer of paint or molded plastic, or an enamel coating that is baked at a very high temperature (900°C).

The leads are usually wire, between 0.6 and 0.8 mm in diameter, that has been tinned for improved solderability.

For high-frequency applications noninductive resistors are used, since their coefficient of induction is only $1/10$ that of standard resistors. The low coefficient is obtained by using two identical parallel windings running in opposite directions, with an ohmic value twice that of the nominal value. Influence of the parasitic inductance is represented by a response curve (see Fig. 4).

2.2.2 Standard Wirewound Resistors

These resistors differ from semiprecision resistors in the following ways:

- nature of the substrate, which is often a fiberglass core
- nature of the coating (cement or plastic)
- technology used for the winding and for adjusting ohmic values, and the method of joining the leads to the resistive element

Maximum allowable temperatures depend on the nature of the coating and can vary between 155 and 275°C.

TABLE 5 Standard Performance of Wirewound Power Resistors

Dissipated Power at 25°C W	Dimensions of Housing Diameter mm	Length mm	Range of Resistance Values Ω	Specific Characteristics
10	9	16	0.01 to 10^4	heat sink
15	10	35	1 to $15 \cdot 10^3$	wire
50	15	50	0.01 to 10^5	heat sink
100	25	125	2 to $82 \cdot 10^3$	wire
300	30	350	4.7 to $220 \cdot 10^3$	wire
300	40	250	0.2 to 33	ribbon
1000	50	375	10 to $47 \cdot 10^3$	wire
1000	50	375	0.39 to 68	ribbon

Power film resistors are also made, whose performance is very different from that of wirewound resistors. They dissipate a few watts of power in a substrate that is often 20 mm long and 5 mm in diameter.

2.2.3 Power Wirewound Resistors

Wirewound resistors were classified above on the basis of their precision and operating stability. We can also use dissipation (see section 1.3 above) as a criterion for resistor selection, sometimes independently of precision.

In this case power is dissipated through a substrate of the appropriate material and size. The wire itself will have a larger cross-section, and in high-power applications flat wound ribbons are often used. In many cases the resistive element is enclosed in a metal box that functions as a heat sink (See Chap. 9, Power Resistors, Fig. 1d). Table 5 lists the performance characteristics of some typical resistors.

3. POTENTIOMETER CHARACTERISTICS

3.1 Taper and Conformity

Taper is the relation between the mechanical position of the wiper and the resistance between the leads a and b (Fig. 5 and Chap. 11, Fig. 1). It is expressed by the ratio V_{ab}/V_{ac}. In most cases the curve is linear. Log normal and inverse, or antilogarithmic, tapers also exist, however. Tapers that

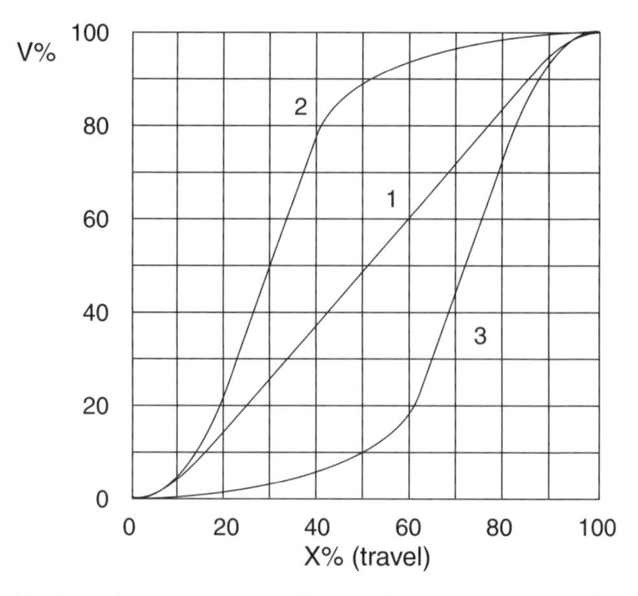

Fig. 5 Tapers in a potentiometer. Percent of voltage vs. percent of travel.

are expressed by more complex relationships (sine, cosine, etc.) can also be designed, especially for precision potentiometers.

Potentiometers are also characterized by conformity—the deviation from the theoretical response. Its full description and physical significance is treated in Chap. 11, Potentiometers.

3.2 Total Mechanical and Electrical Travel

The total mechanical travel is generally limited by two mechanical stops or two positions corresponding to the release of the wiper when it is no longer driven by the control shaft (Fig. 6).

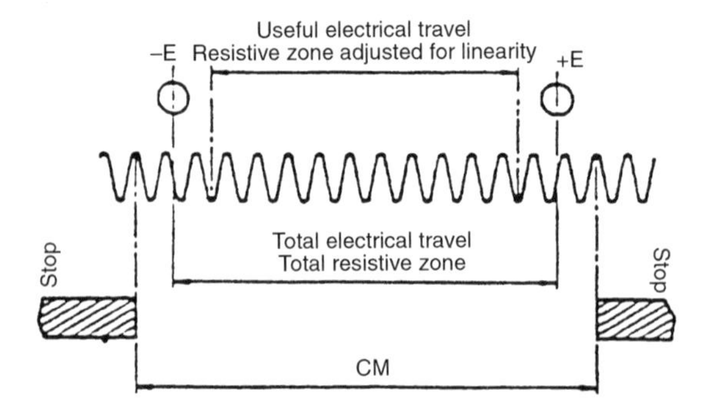

Fig. 6 Mechanical and electrical travel.

Total electrical travel corresponds to the region in which electrical continuity between the wiper and the resistive element is ensured. The useful electrical travel is the region in which the taper is guaranteed.

Some potentiometers do not include stops or clutching mechanisms and are characterized by their total electrical travel and neutral region (where no electrical continuity is ensured).

3.3 Contact Resistance

Contact resistance, denoted as R_c, refers to the region at the point where the wiper hits the resistive element. The value of this contact resistance depends on the type of resistive element and wiper used, the resistivity of the element at the points in contact with the wiper, the contact pressure of the wiper on the resistive element, the size of the contact surfaces, and their condition (Fig. 7).

This contact resistance fluctuates as the wiper is moved across the element.

This variation ΔR_c (and not the average value of R_c) can be the source of disturbance when the potentiometer is used as a rheostat. It has no effect, however, when the potentiometer is used as a voltage divider (providing that the output impedance of the wiper is sufficiently high).

The method of measuring the contact resistance is treated in Chap. 11, Potentiometers.

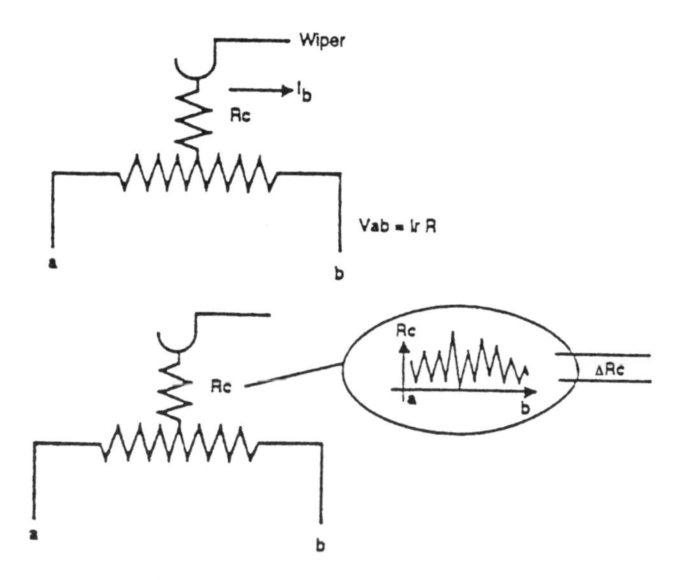

Fig. 7 Contact resistance fluctuation.

3.4 Setting Stability

This test can be applied only to nonwirewound precision potentiometers. It consists of submitting parts to an environmental test and measuring the largest voltage change across a given section of the useful electrical travel. The consistency of the output voltage, expressed as a percentage of the voltage E, is defined as the largest voltage change recorded on the oscilloscope between two points of the useful electrical travel that are no more than 1% from this travel.

4. PRECISION RESISTORS

In general, high-precision resistors exhibit the following performance characteristics:

- Tolerances of ±1% to ±0.001%
- TCR < 5–10 ppm/°C
- TCV < 0.1 μV/V
- Stability $\Delta R/R$ < 500 ppm @ 155°C and 2000 h

Other parameters, such as induction or the relationship to a voltage scale or excess noise, are also used. Noise should be less than −35dB.

Table 6 lists the advantages and disadvantages of the various resistor types.

TABLE 6 Characteristics of Various Precision Resistors

VΩ	<1Ω	<10Ω	10Ω to 100,000Ω	>100,000Ω
Stability ΔR/R < 500 ppm			wirewound foil thin film	
500 < ΔR/R < 1000 ppm	Dale metal strip	Dale film wirewound	wirewound foil thin film cermet	wirewound cermet
ΔR/R > 1000 ppm	Dale metal strip	Dale film wirewound	wirewound foil thin film cermet	wirewound thin film
TCR <5 ppm/°C			wirewound foil thin film	
<20 ppm/°C	Dale metal strip	Dale film wirewound	wirewound foil thin film	wirewound thin film
>20 ppm/°C	Dale metal strip	Dale film wirewound	wirewound foil thin film cermet	thin film cermet
Integration			cermet thin film foil (various)	
Response time $\leq 10^{-9}$	Dale metal strip	Dale film	cermet thin film	cermet
$\leq 10^{-6}$	Dale metal strip	Dale film	cermet thin film foil	cermet
Noise < −35 dB	Dale film	Dale film wirewound	foil thin film	foil thin film

8

Resistive Circuits[1]

1. INTRODUCTION

In the preceding chapters we saw that there is no such thing as a "pure resistor" and that secondary effects arising from the specific mechanisms of conduction or the geometry of resistive components are responsible for deviations or changes in the actual behavior of a resistor from Ohm's law. In the current chapter we will examine and determine the limits of this behavior.

2. PARASITIC RESISTANCE, INSULATION, AND CONNECTIONS

2.1 Influence on Ohmic Values

For a resistor to be used or even measured, it must be connected to the remainder of the circuit or measuring device by means of conductors, which, like the resistor itself, introduce a non-zero ohmic value and self-

[1]Significant portions of this chapter have been taken in their entirety from internal papers prepared by Mr. Cherbuy, a research engineer with the Vishay–SFER Corporation, as well as from application notes on the characteristics of type S-102 resistors, prepared by the Vishay–Foil division.

inductance that must be taken into account if the current varies with time. In the same manner the dielectric medium that protects and insulates the resistor from external effects is characterized by very high, but not infinite, resistance and capacitance that influence the signal.

A resistor placed between the nodes of a network (Fig. 1) can be represented (in DC) by three local resistances: R, the nominal resistance; P, the insulating resistance between the input and output of the nominal resistance, and S, the resistance of the conductors (Fig. 2).

Because of the conservation of current, $I = I_P + I_R$, and the equivalence of the potential differences at the terminals of R and P (with all the resistances being included in P, R, and S, the conductors are considered perfect), we can write:

$$I = I_P + I_R = \frac{U - U_S}{P} + \frac{U - U_S}{R} = U - U_S\left(\frac{1}{P} + \frac{1}{R}\right)$$

$$= U - U_S \cdot \frac{R + P}{PR} \quad (1.1)$$

The same current I also crosses the series resistor S, and the potential between the two nodes is:

$$U = \left(\frac{RP}{R + P} + S\right)I \quad (1.2)$$

The equivalent resistance of the circuit diagram shown in Fig. 2 is:

$$R_e = R - \frac{R^2}{P} \cdot \frac{1}{1 + R/P} + S \quad (1.3)$$

We see that, when expressed in this form, the influence of the insulating resistance P must be such that $P \gg R^2$ to avoid influencing the nominal value of the resistance.

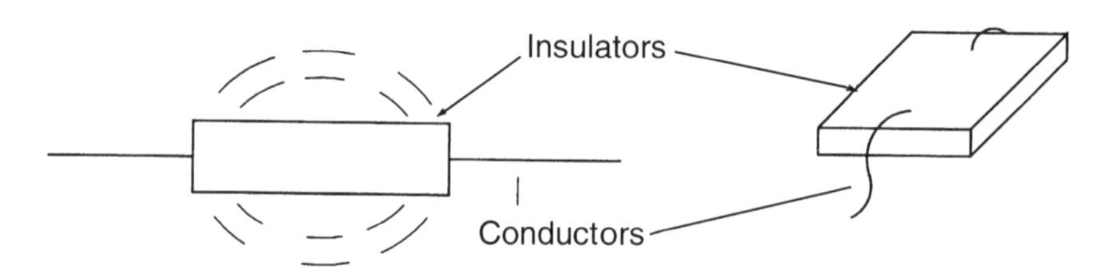

Fig. 1 Conductors and insulators in a resistor.

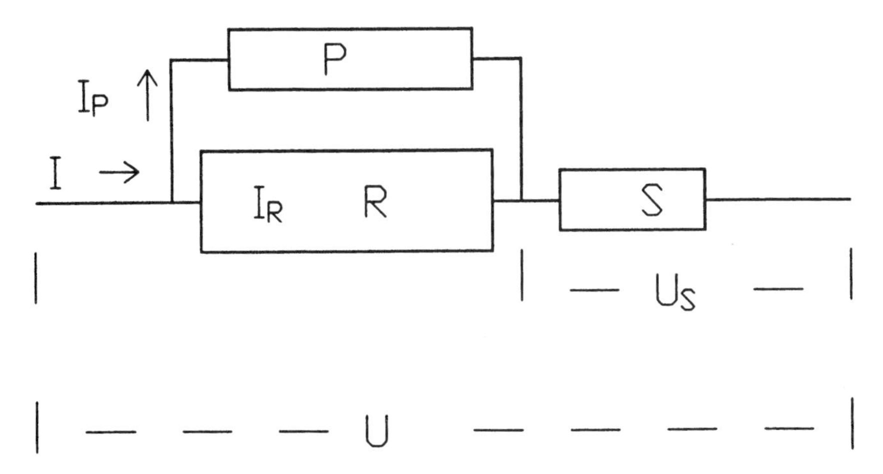

Fig. 2 Electrical schematic of a resistor per Fig. 1.

In practice, the resistance *P* does not function as a discrete resistor connected between *R* and *S* but as one that is distributed over the length of *R*.

If, however, the condition $S \ll R \ll P$ is satisfied, the actual measured resistance is:

$$R_e = R + S - \frac{R^2}{P} = R\left(1 + \frac{S}{R} - \frac{R}{P}\right) \tag{1.4}$$

The orders of magnitude of the resistances *P* and *S* are:

Insulating resistance $P \approx 10^9\ \Omega$

Conductor resistance *S*: $10^{-2}\ \Omega$ to $3 \cdot 10^{-1}\Omega$, depending on the type of conductor or technology used (copper wire, Au or Pd–Ag terminal pads).

We see that measurement errors are minimized when $S/R - R/P = 0$ (Eq. 1.4), that is, when $R = \sqrt{SP} \cong 3k\Omega$.

Mid-range ohmic values (between roughly 10 and 10^6 ohms) are the least sensitive to measurement error. Accurate measurement of very high ohmic values (S/R is negligible, R/P is not) still remains difficult. Four-wire assemblies, however, generally result in correct measurements for low values (R/P is negligible, S/R is not).

Changes in *S* and *P* as a function of their temperature coefficients and the external constraints imposed by the environment will, of course, influence the resistive behavior, whose ohmic value, as seen by the rest of the circuit, is R_e. For the general case, this variation can be written:

$$dR_e = \frac{\partial R_e}{\partial R}dR + \frac{\partial R_e}{\partial P}dP + \frac{\partial R_e}{\partial S}dS \tag{1.5}$$

based on which,

$$\frac{dR_e}{R_e} = \left(\frac{R\partial R_e}{R_e\partial R}\right)\cdot\frac{dR}{R} + \left(\frac{P\partial R_e}{R_e\partial P}\right)\cdot\frac{dP}{P} + \left(\frac{S\partial R_e}{R_e\partial S}\right)\cdot\frac{dS}{S} \tag{1.6a}$$

and

$$\frac{dR_e}{R_e} = \frac{dR}{R}\cdot\left(1 + \frac{\frac{S}{R}\left(\frac{dS/S}{dR/R} - 1\right) + \frac{R}{P}\left(\frac{dP/P}{dR/R} - 1\right)}{1 + S/R + P/R}\right) \tag{1.6b}$$

The terms between parentheses represent the respective values of R, P, and S, for the relative changes measured across R_e. By neglecting the terms S/R and R/P opposite 1 in Eq. 1.6b, we can write:

$$\frac{dR_e}{R_e} = r\left[1 + \frac{S}{R}\left(\frac{s}{r} - 1\right) + \frac{R}{P}\left(\frac{p}{r} - 1\right)\right] \tag{1.7}$$

or

$$\frac{dR_e}{R_e} = r + \frac{S}{R}(s - r) + \frac{R}{P}(p - r) \tag{1.8}$$

in which $r = dR/R$, $s = dS/S$, and $p = dP/P$.

Equation 1.8 can be used to estimate the influence of the changes in P and S on the stability of resistive components. In precision resistors these effects may not be insignificant.

When we analyze the influence of the coefficients of temperature for S and P, it is not uncommon to find insulating materials P with TCR on the order of -10^5 ppm/°C, which, for a $\Delta T°$ of 100°C, results in $p \approx -10$. Equation 1.8 is not applicable for large values of p, but the relative variation of resistance can be deduced from Eq. 1.4 by replacing R/P by R/P_{hitemp} and neglecting R/P_{lowtemp} and S/R. The variation is then equivalent to insertion of insulation's shunting at high temperature, $dR_e/R_e = -R/P_{\text{hitemp}}$. Conductive materials S (Cu or Au) have TCR of +4000 ppm/°C (or 4.10^{-3}/°C), that is, for the same values of $\Delta T°$, $s \approx 4 \cdot 10^{-1}$.

Under these conditions, if the resistive material has been selected so that $r \approx 0$, then $dR_e/R_e \approx 4 \cdot 10^{-1} \cdot S/R - R/P_{\text{hitemp}}$. Depending on the ohmic values chosen, the relative variation in measured resistance can easily reach 10^{-2} (or 1%, for $R = 10\ M\Omega$ and $P_{\text{hitemp}} = 10^9\Omega$) and $4 \cdot 10^{-3}$ (or 0.4%, for $R = 10\Omega$ and $S = 0.1\Omega$), which is far from negligible in the majority of applications. (The contribution from S will have a greater influence on low ohmic values under 100 Ω; the contribution from P will affect high ohmic values.)

The skill of the manufacturer of resistive components consists, therefore, of choosing materials for R, S, and P, such that, depending on the values of R_e, the effects of S and/or P will be minimized.

3. RESISTOR ASSEMBLIES (RATIO, DIVIDERS, BRIDGES, ETC.)

In the previous section we examined the influence on resistor behavior of the undesired, but real, series and parallel parasitic resistances caused by individual elements in the component. In this section we will consider the behavior of a group of interconnected resistors.

3.1 Ratio

In the majority of cases, when modeling an analog circuit, we do not specify individual resistors but the ratio between two resistors. This is particularly true when developing differential amplifiers, for which overall performance is narrowly tied to the performance of resistor ratios at the input and output.

Assume that $\zeta = R_1/R_2$ is a given ratio. The "observed" variation $\delta\zeta$ of this ratio as a function of changes in R_1 and R_2 is such that:

$$\zeta + \delta\zeta = \frac{R_1 + \delta R_1}{R_2 + \delta R_2} = \frac{R_1(1 + r_1)}{R_2(1 + r_2)}$$

based on which

$$\frac{\delta\zeta}{\zeta} = (r_1 - r_2) \cdot \frac{1}{1 + r_2} \tag{1.9}$$

The deviation in the "observed" ratio is, therefore, not linear, contrary to what a simple logarithmic differentiation might imply. When making very precise measurements, the term r_2 may not be negligible to the second order.

The deviation r generally assumes the form $r = \alpha\varepsilon + \beta\varepsilon^2 + \ldots$, in which α, β, etc., are the sensitivities of the system to the constraints ε, ε^2, etc. When considering a thermal constraint, α represents the TCR of the component and $\varepsilon = \Delta T°$, the temperature rise. By neglecting the nonlinear term:

$$\delta\zeta/\zeta = r_1 - r_2 = \alpha_1\varepsilon_1 - \alpha_2\varepsilon_2 \tag{1.10}$$

we can write

$$\alpha_1 = \frac{\alpha_1 + \alpha_2}{2} + \frac{a_1 - \alpha_2}{2} = \overline{\alpha} + \frac{\delta}{2}$$

$$\alpha_2 = \frac{\alpha_1 + \alpha_2}{2} - \frac{a_1 - \alpha_2}{2} = \overline{\alpha} - \frac{\delta}{2}$$

and $\varepsilon_1 = \overline{\varepsilon} + \Delta/2$; $\varepsilon_2 = \overline{\varepsilon} - \Delta/2$, so that

$$\frac{\delta\zeta}{\zeta} = \overline{\alpha}\Delta + \delta\overline{\varepsilon} \tag{1.11}$$

This equation is important in applications and shows that the stability of the ratio depends not only on the product $\delta \cdot \overline{\varepsilon}$, which is the differential sensitivity multiplied by the average constraint, but on the product $\overline{\alpha} \cdot \Delta$ as well (average sensitivity multiplied by the differential constraint). This second factor is generally ignored. It is very important, however, whenever the dissipation between members of the ratio differs. A simple example will demonstrate this:

Assume a ratio consisting of two discrete resistors, $R_1 = 1$ kΩ and $R_2 = 100$ kΩ. The temperature increase for the resistors is approximately 100°C/W, for a current, I, of 1 mA, circulating between them. Then $\varepsilon_1 = \Delta T_1 = 10^{-1}$°C and $\varepsilon_2 = \Delta T_2 = 10$°C.

If $\alpha_1 = \pm 5$ ppm/°C and $\alpha_2 = \pm 7$ ppm/°C (the resistors are high-precision resistors), then $\Delta\zeta/\zeta$ can vary, depending on the respective signs of α_1 and α_2, from −70 to +70 ppm, which may be too high for certain measurement systems. When discrete resistors are involved (film or wire-wound), we see the obvious advantage in choosing components not only with a TCR differential that is as close to zero as possible ("matching") but also with absolute TCR values that are very small.

When assembly and dissipation conditions for circuits (integrated thin-film resistors on a single substrate, for example) are such that values of ε are closely matched ($\varepsilon_1 \approx \varepsilon_2$), then $\Delta \approx 0$ and only the difference between TCR values or the differences among the mechanical and thermomechanical constraints are significant.

3.2 Voltage Divider (Potentiometer, Half-bridge, etc.)

Here we consider the variation of the output signal $S = V_s/V_T$ (Fig. 3):

$$S = \frac{R_2}{R_1 + R_2} = \frac{1}{1 + \zeta} \tag{1.12}$$

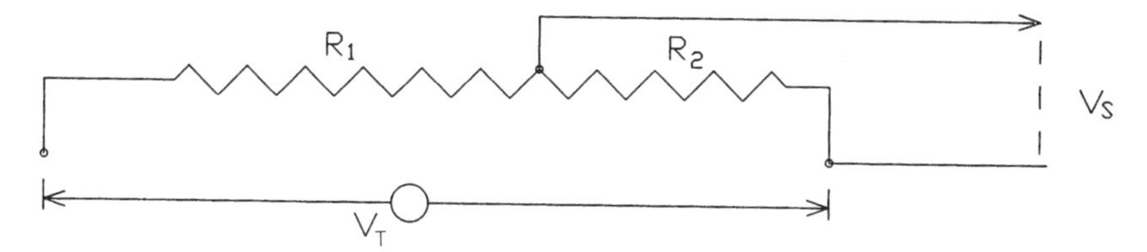

Fig. 3 Schematic of a voltage divider.

and

$$\frac{\delta S}{S} = -\frac{\delta\zeta/\zeta}{1 + 1/\zeta} = \frac{r_1 - r_2}{(1 + r_2)(1 + 1/\zeta)} \tag{1.13}$$

This enables us to determine the deviation of the signal S as a function of the deviation of the ratio of the two resistances.

The previous example, which gave an error of 70 ppm for the ratio, can thus be used to determine that the half-bridge used as a voltage divider will have an error of approximately 8 ppm caused by the absolute and relative TCR values.

3.3 Wheatstone Bridge

In the case of a Wheatstone bridge (Fig. 4), there are two dividers fed by the same voltage V_T. We can measure the signal:

$$S_W = V_{S2}/V_T - V_{S1}/V_T = V_S/V_T$$

or

$$S_W = \frac{1}{1 + \zeta} - \frac{1}{1 + \xi}$$

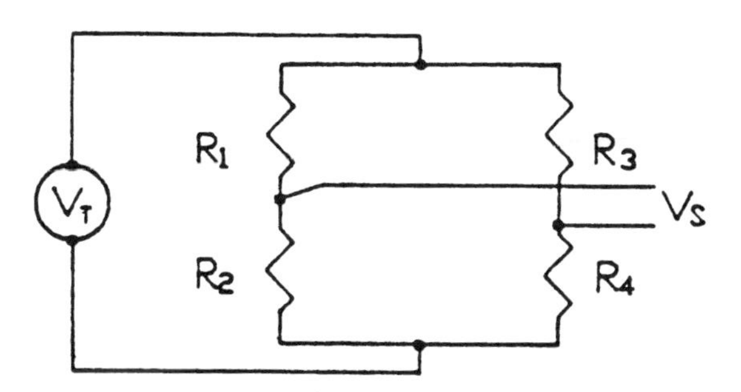

Fig. 4 Schematic of a Wheatstone bridge.

The error δS "observed" when the resistances R_i vary, equals:

$$\delta S_W = -\frac{\zeta(r_1 - r_2)}{(1 + r_2)(1 + \zeta)^2} + \frac{\xi(r_3 - r_4)}{(1 + r_4)(1 + \xi)^2} \tag{1.14}$$

The Wheatstone bridge is traditionally used to compare a resistance (R_1, for example) to a standard R_2 (either constant or a decade box). In this case we do not take the deviation into account, $\Delta S = 0$, and when the bridge is balanced ($S_W = 0$), we get:

$$R_1 = \xi \cdot R_2 \tag{1.15}$$

ξ is called the "arm ratio" of the bridge.

3.3.1 Measurement of the Deviation of a Resistor

When r_2, r_3, and r_4 are negligible, we measure the variation in r_1 of the resistance R_1 by means of the deviation of the signal:

$$\delta S_W = -\frac{\zeta r_1}{(1 + \zeta)^2} \tag{1.16}$$

3.3.2 Use of a Bridge in a Resistance Strain Gage

All resistors used as strain gages have the same initial value ($\zeta \cong \xi \cong 1$) and vary by no more than a few 10^{-3} ohm (negligible when compared to $1 + \zeta$ or $1 + \xi$). In this case the variation of the signal can be written:

$$\delta S = \frac{1}{4}[(r_1 + r_3) - (r_2 + r_4) + [r_2^2 + r_4^2 - r_1 r_2 - r_3 r_4]] \tag{1.17}$$

4. FREQUENCY RESPONSE OF RESISTORS

Because of their shape, resistors have both capacitance and self-inductance associated with purely resistive functions. A wirewound resistor will have a coefficient of self-induction that is sometimes considerable. The proximity of wires or ribbons can also create parasitic capacitance.

At very high frequencies (above 100 MHz), the losses in dielectric materials, together with the "skin effect" described in Chap. 1, are no longer negligible. All these factors contribute to modify the response of the "resistive component" when it is subject to voltages or currents that change over time.

By neglecting UHF effects (losses in the dielectric) and for frequencies between a few Hz and 100 MHz, the resistor can be represented by the equivalent circuit diagram[2] shown in Fig. 5.

The impedance of this circuit is given by the following equation:

$$Z(p) = \frac{p + R/L}{C(p^2 + p \cdot R/L + 1/LC)} \tag{1.18}$$

If $p = j\omega$ (for example, an AC voltage of frequency $\omega = 2\pi f$), it is easy to show that:

$$Z(\omega) = R_Z + j\omega \cdot L_Z \tag{1.19}$$

in which

$$R_Z = \frac{R}{(1 - \omega^2 LC)^2 + (\omega RC)^2} \tag{1.20a}$$

$$L_Z = \frac{L(1 - \omega^2 LC) - R^2 C}{(1 - \omega^2 LC)^2 + (\omega RC)^2} \tag{1.20b}$$

The admittance, Y, of the circuit is:

$$Y = 1/Z = 1/R_y + j\omega C_Y \tag{1.21}$$

in which

$$R_Y = R[1 + \omega^2 L^2/R^2] \tag{1.22}$$

$$C_Y = C - \frac{1}{L\omega^2[1 + \omega^{-2}(R/L)^2]} \tag{1.23}$$

The voltage response to a range of current pulses at $t = 0$ $(1/p)$ is given by the transform in the time domain of equation $E(p) = Z(p)/(p)$. The shape

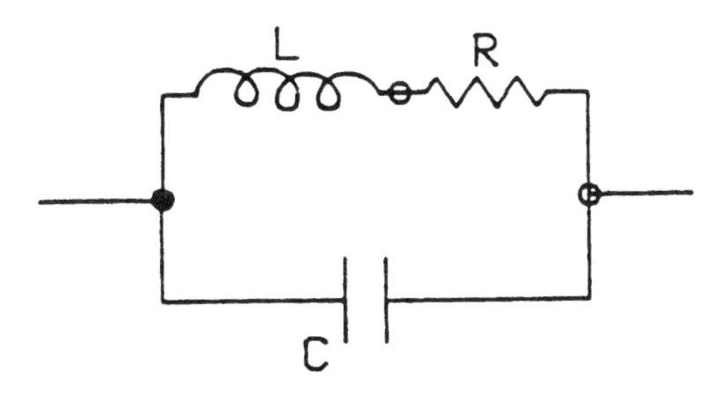

Fig. 5 Schematic of a resistor with its inductance and capacitance.

[2]Vishay. Personal communication and technical papers.

of the output voltage $E(t)$, calculated from the Laplacian transformation, depends on the relative values of R, L, and C.[3] If $R^2/L^2 \gg 4/LC$, then:

$$E(t) = R[1 - e^{-t/RC}] \tag{1.24}$$

If $R^2/L^2 < 4/LC$, then:

$$E(t) = R\left[1 - \frac{1}{RC\omega_0} e^{-(R/2L)t} \cos(\omega_0 t + \alpha)\right] \tag{1.25}$$

in which

$$tg(\alpha) = \frac{[R^2/2L^2 - 1/LC]}{\omega_0 R/L} \quad \text{and} \quad \omega_0 = \frac{1}{2}[4/LC - R^2/L^2]^{1/2} \tag{1.26}$$

If $4/LC \gg R^2/L^2$, then:

$$E(t) = R\left[1 - \frac{1}{R}\sqrt{L/C} \cdot e^{-(R/2L)t} \cos(\omega_0 t + \alpha)\right] \tag{1.27}$$

in which

$$tg\ \alpha = -\frac{1}{R}\sqrt{L/C} \quad \text{and} \quad \omega_0 = 1/\sqrt{LC}. \tag{1.28}$$

For example, the frequency response of type S-102 high-precision resistors is approximately:[4]

$C_Y \cong 0.5pF$ ($R = 30\ k\Omega$) and $L_Z \cong 80.10^{-3}\ \mu H$ ($R = 400\ \Omega$) at 100 MHz.

Of course, these values depend on the geometry of the resistors (number of strands, size of the strands, etc.) and consequently on the ohmic value. This is illustrated in Table 1, showing the transition time of Vishay S-102 resistors as a function of their ohmic value, when subject to a current step.

Figure 6 shows the shape of the response curve of resistance as a function of ohmic value.

Obviously the analysis discussed in this section is valid for resistive networks during operation when subject to voltages or currents that vary with

[3]Vishay. Personal communication and technical papers.

[4]Vishay. Technical documentation.

TABLE 1 Transition Time of Vishay Resistors

Resistance	Response	Transition Time
10 Ω	Oscillating	16.0 ns
100 Ω	Oscillating	1.60 ns
800 Ω	Dampened	0.20 ns
1 kΩ	Exponential	0.16 ns
5 kΩ	Exponential	5.00 ns
10 kΩ	Exponential	15.0 ns
100 kΩ	Exponential	40.0 ns

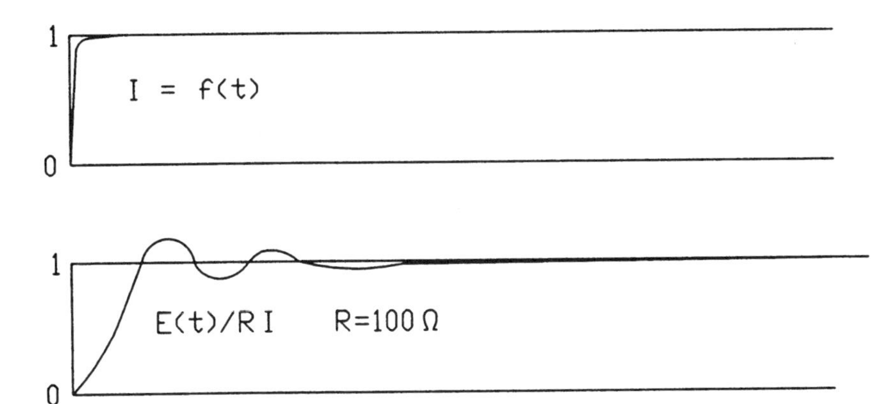

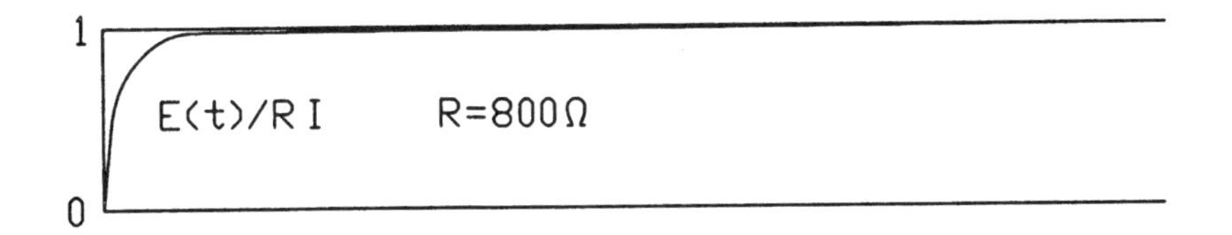

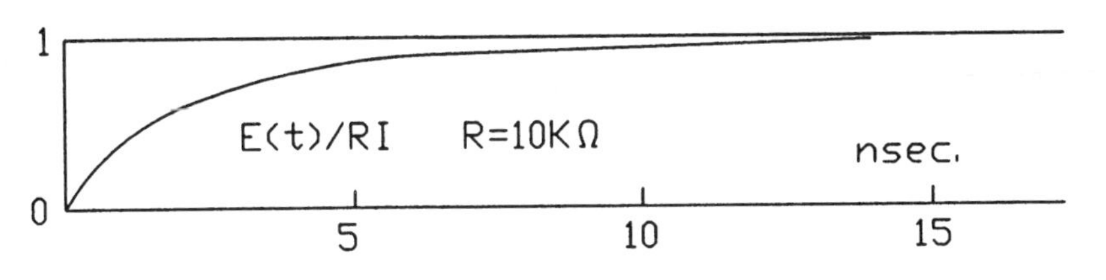

Fig. 6 Response curves of resistors.

time. In this case the values of r_i are the relative responses of each resistive element as a function of the frequency and other constraints. As soon as several elements, j, representing different sources of constraint, are applied simultaneously to the resistances, then, applying standard procedures for dealing with multiple sources of error, we define r_i by statistical equation:

$$r_i = \left(\sum^{j} r_{ij}^2\right)^{1/2} \tag{1.29}$$

9

Power and Wirewound Resistors

1. INTRODUCTION

There are four principal types of wirewound resistors:

- Axial or radial lead resistors (Fig. 1a)
- Surface-mounted resistors (Figs. 1b, 1c)
- Power resistors, still referred to as housed resistors (HR) (Fig. 1d)
- High-power tubular resistors (Fig. 1e)

Axial lead resistors consist of a winding of resistive wire on a cylindrical ceramic substrate. The current or voltage leads are made of metal wire (generally copper plated with a tin-lead alloy) welded at the ends of the cylinder to metal caps to which the ends of the resistive wire are also welded. Depending on the application, the leads are welded either axially (parallel to the axis of the cylinder) or radially (perpendicular to the axis of the cylinder). The body of the resistor is covered with a protective enamel or mineral-filled silicone coating.

Power resistors are made in the same way as radial lead resistors. However, they are then placed in a metal container that functions like a heat sink, and through which they dissipate a considerable amount of power.

Their range extends from a few ohms to approximately 10^4 ohms, with a TCR between ±20 to 500 ppm/°C, depending primarily on ohmic

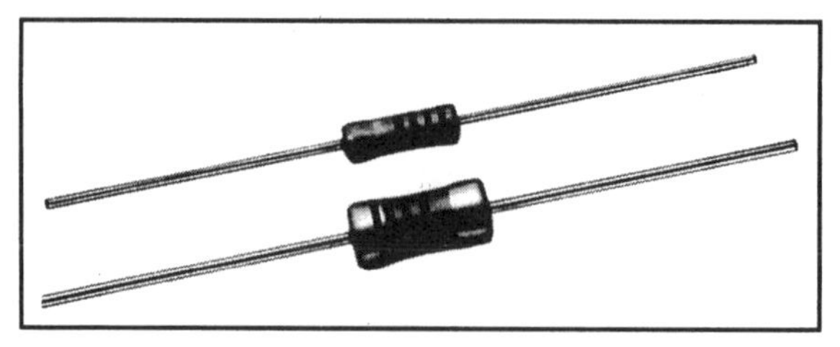

(a) axial, low-power resistors

(b) surface-mounted molded resistors

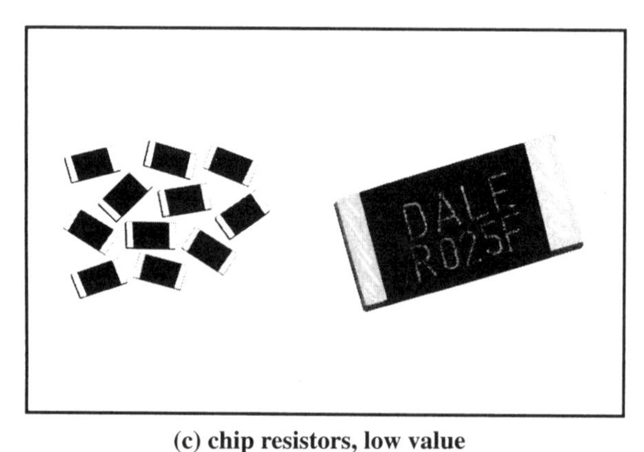

(c) chip resistors, low value

Fig. 1 Families of wirewound resistors

value. Figure 1 presents some of the standard configurations and shapes for these types of resistors.

Nonwirewound high power surface-mounted resistors, specially designed in order to dissipate 2 to 5 times more electrical power than the HR type housed ones, are described in Section 5.

2. LOW-POWER RESISTORS

The expression "low-power resistor" is somewhat arbitrary. In reality wirewound resistors dissipate much more power than other types of resistors. In this case the dissipated power is between 0.125 W and 10 W.

(d) housed power resistor

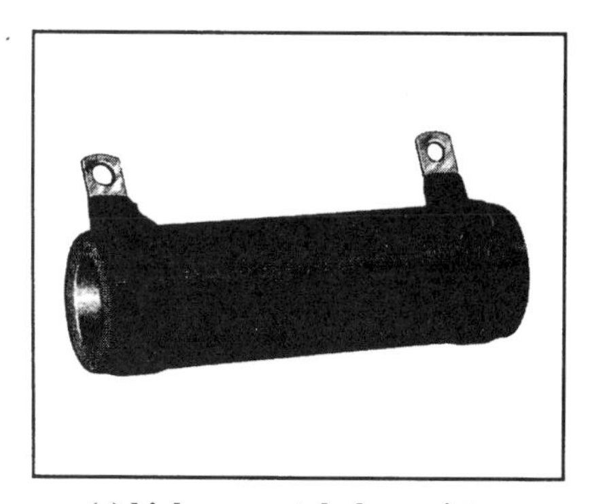

(e) high-power tubular resistor

Fig. 1 (*continued*)

Figure 2 shows the configuration of some typical low-power resistors. These components are manufactured as follows:

1. The ceramic is washed in the appropriate solvents, followed by treatment in deionized water to eliminate any organic impurities. It is then heated to approximately 450°C to eliminate any traces of water.
2. Automated machinery is used to attach the caps.
3. Winding. This operation is generally carried out on automated machinery capable of high output rates, roughly 2,000 to 4,000 parts per hour. Figure 3 is an illustration of one such machine.

MODEL G

APPLICABLE MIL-SPECIFICATIONS

MIL-R-26E: This is a military specification designed especially for precision and non-precision power wirewound resistors. The G models meet the requirements of this specification as well as the older MIL-R-26C and MIL-R-23379 specifications.

SPECIAL MODIFICATIONS

1. Terminals can be supplied in any commercial material with several type finishes.
2. Terminal lengths and diameters can be varied.
3. Various elements available for special TC.
4. Special configuration available on request.
5. Tolerances available to ± .01% on most types.
6. Special matching available (TC and tolerance).

GN - NON-INDUCTIVE

Models of equivalent physical and electrical specifications are available with non-inductive (Aryton-Perry) winding. They are identified by adding the letter N to the letter G in the part number (GN-5, for example). Three conditions apply:

1. For GN Types, divide maximum resistance values by two.
2. For GN Types, multiply maximum working voltage by .707.
3. For GN Types, maximum weights may slightly exceed those shown on low values.

GN-1-80	GN-5
GN-2	GN-5C*
GN-3-80	GN-10

* Body O.D. may exceed that of the G-5C by .010" [.254mm].

ENVIRONMENTAL PERFORMANCE *

TEST	MIL-R-26E REQUIREMENT	DALE MAXIMUM
Moisture Resistance	± (.2% + .05Ω) ΔR	± (.2% + .05Ω) ΔR
Load Life	± (.5% + .05Ω) ΔR	± (.5% + .05Ω) ΔR
Temperature Coefficient	30 - 90PPM/°C Max.	See Electrical Specifications
Thermal Shock	± (.2% + .05Ω) ΔR	± (.2% + .05Ω) ΔR
Short Time Overload	± (.2% + .05Ω) ΔR	± (.2% + .05Ω) ΔR
Dielectric	± (.1% + .05Ω) ΔR	± (.1% + .05Ω) ΔR
Low Temperature Storage	± (.2% + .05Ω) ΔR	± (.2% + .05Ω) ΔR
High Temperature Exposure	± (.5% + .05Ω) ΔR	± (.5% + .05Ω) ΔR
Shock	± (.1% + .05Ω) ΔR	± (.1% + .05Ω) ΔR
Vibration	± (.1% + .05Ω) ΔR	± (.1% + .05Ω) ΔR
Terminal Strength	± (.1% + .05Ω) ΔR	± (.1% + .05Ω) ΔR

* All ΔR figures shown are maximum, based on units with an initial tolerance of ± 1% and maximum operating temperature of + 275°C.

DERATING

Dale® G style resistors have an operating temperature range of - 55°C to + 275°C (Characteristic U) or - 55°C to + 350°C (Characteristic V). See Electrical Specifications table. They must be derated at high ambient temperatures according to the curves below.

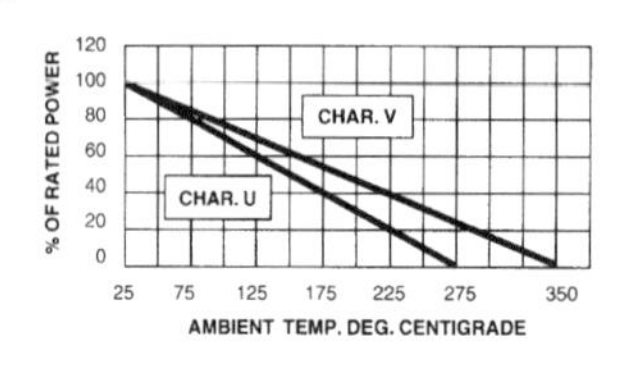

CHARACTERISTIC U: Resistors are available in any tolerance.

CHARACTERISTIC V: Resistors are available in ± 3% and ± 5% tolerance.

POWER RATING

Power ratings of Dale® G resistors are 1.4 to 4 times higher than those of conventional wirewound resistors of equivalent size. At the higher ratings, Dale® G resistors will meet the same environmental and life stability requirements of the lower rated conventional resistors.

CHARACTERISTIC U:

1. + 275°C maximum hotspot temperature.
2. .5% maximum ΔR in 2000 hour load life.

CHARACTERISTIC V:

1. + 350°C maximum hotspot temperature.
2. 3% maximum ΔR in 2000 hour load life.

PART MARKING

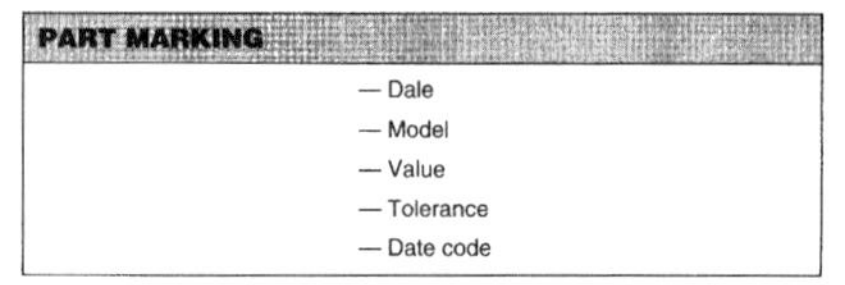

HOW TO ORDER

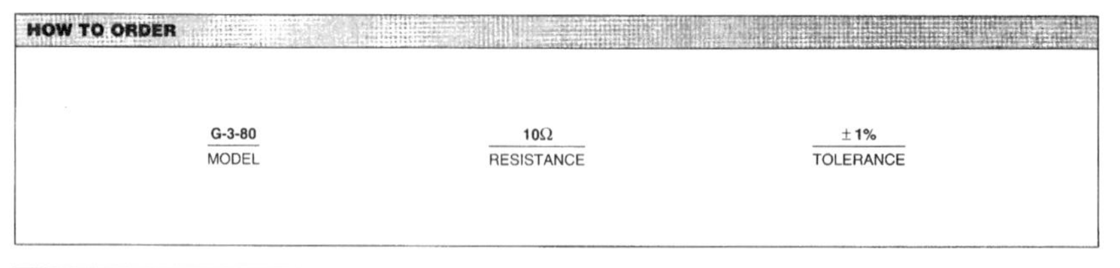

Fig. 2 Axial resistors, 1/8 to 10W (Dale, type RS).

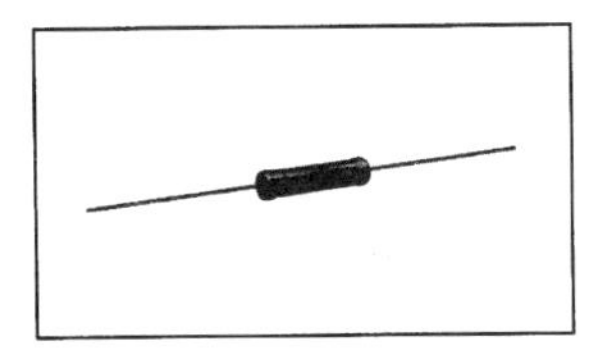

FEATURES

- Complete welded construction
- High-temperature silicone coating
- Meets applicable requirements of MIL-R-26
- Available in non-inductive styles (Type NS) with Aryton-Perry winding for lowest reactive components
- Over 44 million unit-hours of testing with no catastrophic failures have proven failure rate of less than 0.0066% per 1000 hours (at 60% confidence) with full rated power at + 25°C. A failure is defined as ± 1% resistance change.

Fig. 2 (*continued*)

Fig. 3 High volume wire winding machine (Dale).

Some custom resistors are wound on manual machines. The tolerance is controlled by the length of wirewound around the substrate.

4. The lead wires are welded to the caps by means of automatic electric welding machines. Wires are generally made of tinned copper.
5. The resistor is then protected with several layers of silicone-based resin or, depending on the application, with low-temperature enamel. Protective coatings are applied by painting, molding, or dipping the resistor in a fluidized bed.
6. The resistor undergoes a final measurement, and is marked with its value in ohms and its tolerance; samples are taken for quality

control testing or, for more demanding applications,[1] each part is individually tested.

7. Depending on the intended application, the component, in addition to ordinary quality-control operations, may be subject to accelerated aging and repeated power overload tests for several hours while in operation. These tests are designed to eliminate components with a high probability of failure during the first few hours of operation.

2.1 Electrical and Mechanical Characteristics

The electrical and mechanical characteristics of wirewound resistors are shown in Tables 1 and 2 (courtesy of Dale). Dissipated power depends on the dimensions or volume of the component. Through the use of statistical regression, we can determine a (very) approximate relation between dissipated power (in watts) and the weight in grams for axial lead resistors:

$$W = 0.3 + 1.5p - 0.10p^2 \tag{1.1}$$

in which W is the power dissipated at an external temperature of 25°C and p is the weight of the component (in grams). Table 1 shows the characteristics of some axial resistors. Table 2 shows the mechanical characteristics for these same resistors.

The limit of dissipated power depends on the temperature reached by the resistive wire, whose average temperature should not exceed 350–360°C. The wires used (constantan or Ni-Cr) in wirewound resistors are drawn, and during preparation the crystal grains of the polycrystalline ingot are deformed and stretched in the direction of drawing. Metallurgists refer to this process as *strain hardening*. If the wire is heated to rather high temperatures, the mechanisms of diffusion enable the alloy to mature toward more favorable energy states, reducing the grain boundary surface area per unit volume. No phase change occurs during this process, and the face-centered cubic structure of constantan and Ni-Cr remains after recrystallization. On the other hand, the temperature of recrystallization is

[1]Samples are usually taken for quality control procedures. A certain number of components are removed from the fabrication line or prior to shipment and tested, following the standardized procedures, to determine the maximum number of defective parts. For more rigorous applications, however, in which a breakdown from a defective component is unacceptable (military or space applications, for example) each manufactured component intended for shipment is tested. Readers interested in these issues can refer to Appendices 2 and 3.

TABLE 1 Electrical Characteristics of WW Resistors, Dale Type RS

STANDARD ELECTRICAL SPECIFICATIONS

DALE MODEL	MIL-R-26 TYPE	DALE RATING		RESISTANCE RANGE (Ohms) MIL. RANGE SHOWN IN BOLD FACE				MAXIMUM* WORKING VOLTAGE		WEIGHT (Grams)
		U ±.05% thru ±5%	V ±3% & ±5%	±.05%	±.1%	±.25%	±.5%, ±1% ±3%, ±5%	U	V	
RS-1/8	—	.125 W	—	—	—	—	.1 - 1.4k	8.5	—	.15
RS-1/4	—	.4 W	—	1 - 1k	.499 - 1k	.499 - 3.4k	.1 - 3.4k	20	—	.21
RS-1/2	—	.75 W	—	1 - 1.3k	.499 - 1.3k	.499 - 4.9k	.1 - 4.9k	29	—	.23
RS-1A	—	1.0 W	—	1 - 2.74k	.499 - 2.74k	.499 - 10.4k	.1 - 10.4k	52	—	.34
RS-1A-300	**RW70**	**1.0 W**	—	—	**.499 - 2.74k**	**.499 - 2.74k**	**.1 - 2.74k**	—	—	—
RS-1M	—	1.0W	—	1 - 1.32k	.499 - 1.67k	.499 - 6.85k	.1 - 6.85k	41	—	.30
RS-2	—	4.0 W	5.5 W	.499 - 12.7k	.499 - 12.7k	.1 - 47.1k	.1 - 47.1k	210	250	2.10
RS-2M	—	3.0W	—	.499 - 4.49k	.499 - 4.49k	.1 - 18.74k	.1 - 18.74k	95	—	.65
RS-2B	—	3.0 W	3.75 W	.499 - 6.5k	.499 - 6.5k	.1 - 24.5k	.1 - 24.5k	140	157	.70
RS-2B-300	**RW79**	**3.0 W**	—	—	**.499 - 6.49k**	**.1 - 6.49k**	**.1 - 6.49k**	—	—	—
RS-2C	—	2.5 W	3.25 W	.499 - 8.6k	.499 - 8.6k	.1 - 32.3k	.1 - 32.3k	138	157	1.6
RS-2C-17	—	2.5 W	3.25 W	.499 - 6.8k	.499 - 8.6k	.1 - 32.3k	.1 - 32.3k	138	157	1.6
RS-2C-23**	**RW69**	**2.5 W**	**3.25 W**	—	—	—	**.1 - 2.0k**	**130**	**150**	**1.6**
RS-5	—	5.0 W	6.5 W	.499 - 25.7k	.499 - 25.7k	.1 - 95.2k	.1 - 95.2k	360	410	4.2
RS-5-69	**RW74**	**5.0 W**	—	—	**.499 - 24.3k**	**.1 - 24.3k**	**.1 - 24.3k**	—	—	—
RS-5-70**	**RW67**	**5.0 W**	**6.5 W**	—	—	—	**.1 - 8.5k**	**320**	**365**	**4.2**
RS-7	—	7.0 W	9.0 W	.499 - 41.4k	.499 - 41.4k	.1 - 154k	.1 - 154k	504	576	4.7
RS-10	—	10.0 W	13.0 W	.499 - 73.4k	.499 - 73.4k	.1 - 273k	.1 - 273k	858	978	9.0
RS-10-38	**RW78**	**10.0 W**	—	—	**.499 - 71.5k**	**.1 - 71.5k**	**.1 - 71.5k**	—	—	—
RS-10-39**	**RW68**	**10.0 W**	**13.0 W**	—	—	—	**.1 - 20k**	**765**	**875**	**9.0**

*Maximum working voltage determined at .0008" [.20mm] diameter wire resistance values.
**Standard tolerance is ± 5% .1 ohm and above, ± 10% below 1 ohm. ± 3% available.
Note: Shaded area indicates most popular models.

not constant (as is the Curie point of steel, for example) but depends on the metallurgical history of the alloy and can vary from one spool of wire to another. A temperature that has no apparent effect on one spool of wire will produce very rapid expansion in another spool. Modification of the average grain size can have two significant consequences:

1. Mechanically, the wire will become weaker. This is particularly dangerous for very fine wires (diameters of 10–20 μm) in which the increased grain size means that the grains will occupy a significant portion of the wire cross-section.

2. Electrically, this increase will result in a decrease in the density of grain boundaries, which will lead to an irreversible negative deviation in resistivity (as high as 10–15%) and, through the application of Matthiessen's rule, a corresponding positive deviation in the TCR (see Chap. 3). Figure 4 shows the temperature of the resistor's surface compared to the dissipated power for Dale type LVR resistors.

The values for dissipated power are given in the manufacturers' catalogs at an ambient temperature of 25°C. When the ambient temperature is higher, the nominal power dissipated obeys a relation known as a "derating curve," shown in Fig. 5 for Dale RS type resistors. This curve takes into account the fact that the temperature of the resistive wire cannot exceed

TABLE 2 Mechanical Characteristics of WW Resistors, Dale Type RS

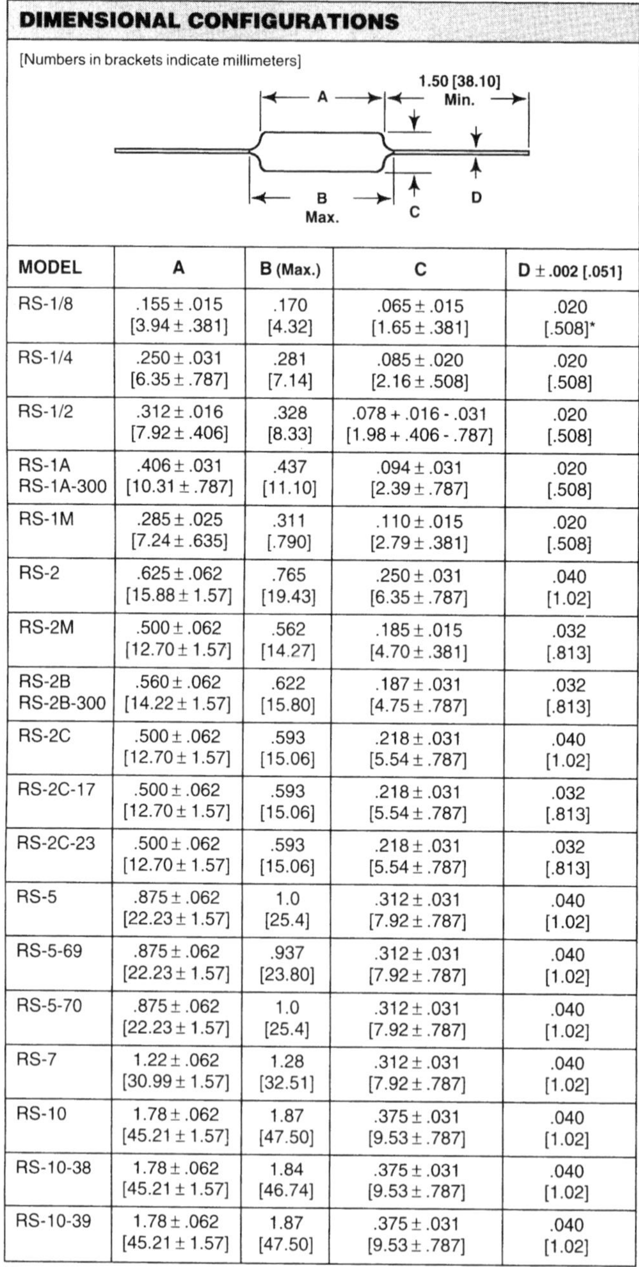

MODEL RS

DIMENSIONAL CONFIGURATIONS

[Numbers in brackets indicate millimeters]

MODEL	A	B (Max.)	C	D ± .002 [.051]
RS-1/8	.155 ± .015 [3.94 ± .381]	.170 [4.32]	.065 ± .015 [1.65 ± .381]	.020 [.508]*
RS-1/4	.250 ± .031 [6.35 ± .787]	.281 [7.14]	.085 ± .020 [2.16 ± .508]	.020 [.508]
RS-1/2	.312 ± .016 [7.92 ± .406]	.328 [8.33]	.078 + .016 - .031 [1.98 + .406 - .787]	.020 [.508]
RS-1A RS-1A-300	.406 ± .031 [10.31 ± .787]	.437 [11.10]	.094 ± .031 [2.39 ± .787]	.020 [.508]
RS-1M	.285 ± .025 [7.24 ± .635]	.311 [.790]	.110 ± .015 [2.79 ± .381]	.020 [.508]
RS-2	.625 ± .062 [15.88 ± 1.57]	.765 [19.43]	.250 ± .031 [6.35 ± .787]	.040 [1.02]
RS-2M	.500 ± .062 [12.70 ± 1.57]	.562 [14.27]	.185 ± .015 [4.70 ± .381]	.032 [.813]
RS-2B RS-2B-300	.560 ± .062 [14.22 ± 1.57]	.622 [15.80]	.187 ± .031 [4.75 ± .787]	.032 [.813]
RS-2C	.500 ± .062 [12.70 ± 1.57]	.593 [15.06]	.218 ± .031 [5.54 ± .787]	.040 [1.02]
RS-2C-17	.500 ± .062 [12.70 ± 1.57]	.593 [15.06]	.218 ± .031 [5.54 ± .787]	.032 [.813]
RS-2C-23	.500 ± .062 [12.70 ± 1.57]	.593 [15.06]	.218 ± .031 [5.54 ± .787]	.032 [.813]
RS-5	.875 ± .062 [22.23 ± 1.57]	1.0 [25.4]	.312 ± .031 [7.92 ± .787]	.040 [1.02]
RS-5-69	.875 ± .062 [22.23 ± 1.57]	.937 [23.80]	.312 ± .031 [7.92 ± .787]	.040 [1.02]
RS-5-70	.875 ± .062 [22.23 ± 1.57]	1.0 [25.4]	.312 ± .031 [7.92 ± .787]	.040 [1.02]
RS-7	1.22 ± .062 [30.99 ± 1.57]	1.28 [32.51]	.312 ± .031 [7.92 ± .787]	.040 [1.02]
RS-10	1.78 ± .062 [45.21 ± 1.57]	1.87 [47.50]	.375 ± .031 [9.53 ± .787]	.040 [1.02]
RS-10-38	1.78 ± .062 [45.21 ± 1.57]	1.84 [46.74]	.375 ± .031 [9.53 ± .787]	.040 [1.02]
RS-10-39	1.78 ± .062 [45.21 ± 1.57]	1.87 [47.50]	.375 ± .031 [9.53 ± .787]	.040 [1.02]

Note: RS-1/8 terminal length will be 1.0" [25.4mm] minimum.

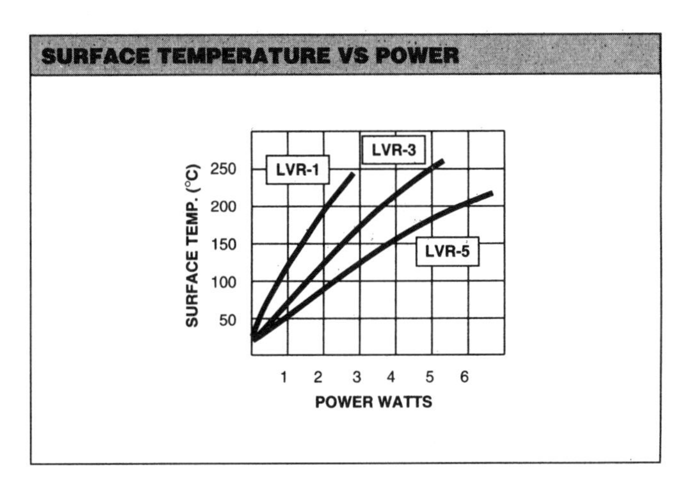

Fig. 4 Surface temperature vs. power on LVR resistors (Dale).

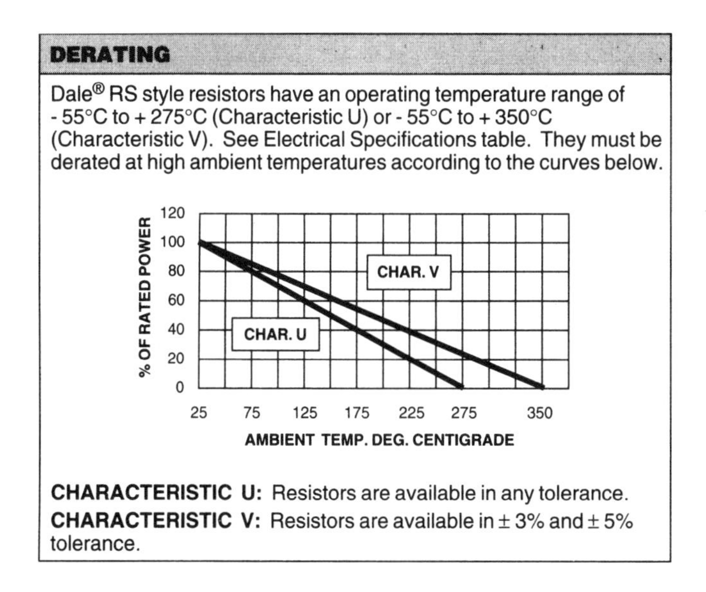

Fig. 5 Derating of applied power in model RS WW resistors (Dale).

360°C. If the ambient temperature reaches this value, the dissipated power in watts will be equal to or near zero.

The method of manufacturing wirewound resistors introduces high self-inductances, with values of approximately 10^{-2} μH, depending on their ohmic value. These resistors cannot be used in high-frequency applications involving frequencies greater than 1–2 MHz.

3. HIGH-POWER RESISTORS

High-power resistors are used, as the name indicates, for applications involving strong currents, where considerable power is involved. Their basic structure does not differ appreciably from that of low-power resistors—a resistive wire is wound around a ceramic substrate. This element is then encased in a metal heat sink that increases the overall mass and diffusion surface for heat dissipation.

Figure 6 illustrates the method of constructing such resistors. The fabrication method is nearly identical to that used for low-power resistors. But instead of using polymer resins for protection, the resistor is coated with a silicone-based lacquer and placed in the cylindrical opening at the center of the heat sink. The source of the greatest thermal resistance in the component is obviously the resistor-heat sink bond, for which lacquer is used. Its coefficient of heat transfer is an order of magnitude lower than the other materials used to manufacture the resistor. This is compensated for by the surface of the heat sink and by the greater diameter of the lead wire, which also acts as a heat sink.

Nevertheless, the power rating is approximately an order of magnitude higher than that of low-power resistors, as shown in Tables 3 and 4, which are supplied by Dale.

Naturally, the derating curves mentioned above also apply to high-power resistive components. At 360°C the dissipated power should be zero to prevent damage to the resistor wire.

Fig. 6 Housed power resistor.

TABLE 3 Specifications of Housed Resistors (Dale)

STANDARD ELECTRICAL SPECIFICATIONS

DALE MODEL	MIL-R-39009 TYPE	POWER RATING (Watts)		MILITARY RESISTANCE RANGE (Ohms) ±1%	MAX. WORKING VOLTAGE	MAX. WEIGHT (Grams)	MIL-R-39009 STANDARD TEMPERATURE COEFFICIENT VALUE RANGES (Ohms)		
		MOUNTED	FREE AIR				±100PPM/°C	±50PPM/°C	±30PPM/°C
ENH-5	RER40	5	3	1 - 1.65k	128.9	3.3	—	1 - 19.9	20 - 1.65k
ENH-10	RER45	10	6	1 - 2.8k	190.0	8.8	—	1 - 19.9	20 - 2.8k
ENH-25	RER50	20	8	1 - 6.04k	390.0	16.5	—	1 - 19.9	20 - 6.04k
ENH-50	RER55	30	10	1 - 19.6k	890.0	35.0	—	1 - 19.9	20 - 19.6k
ERH-5	RER60	5	3	0.10 - 3.32k	160.0	3.0	.1 - .99	1 - 19.9	20 - 3.32k
ERH-10	RER65	10	6	0.10 - 5.62k	265.0	6.0	.1 - .99	1 - 19.9	20 - 5.62k
ERH-25	RER70	20	8	0.10 - 12.1k	550.0	13.0	.1 - .99	1 - 19.9	20 - 12.1k
ERH-50	RER75	30	10	0.10 - 39.2k	1250.0	28.0	.1 - .99	1 - 19.9	20 - 39.2k

Note: All resistance ranges shown conform to military specifications unless otherwise indicated.

TABLE 4 Dimensions of Housed Resistors

MODELS ERH and ENH

DIMENSIONAL CONFIGURATIONS [Numbers in brackets indicate millimeters]

ERH-5, ENH-5 through ERH-50, ENH-50

MODEL	A	B	C	D	E	F	G	H	J	K	L	M	N	P
ERH-5 ENH-5	.444 ±.005 [11.28 ±.127]	.490 ±.005 [12.45 ±.127]	.600 ±.031 [15.24 ±.787]	1.125 ±.062 [28.58 ±1.57]	.334 ±.015 [8.48 ±.381]	.646 ±.015 [16.41 ±.381]	.320 ±.015 [8.13 ±.381]	.065 ±.010 [1.65 ±.254]	.133 ±.010 [3.38 ±.254]	.078 ±.010 [1.98 ±.254]	.093 ±.005 [2.36 ±.127]	.078 ±.015 [1.98 ±.381]	.050 ±.005 [1.27 ±.127]	.266 ±.062 [6.76 ±1.57]
ERH-10 ENH-10	.562 ±.005 [14.27 ±.127]	.625 ±.005 [15.88 ±.127]	.750 ±.031 [19.05 ±.787]	1.375 ±.062 [34.93 ±1.57]	.420 ±.015 [10.67 ±.381]	.800 ±.015 [20.32 ±.381]	.390 ±.015 [9.91 ±.381]	.075 ±.010 [1.90 ±.254]	.165 ±.010 [4.19 ±.254]	.093 ±.010 [2.36 ±.254]	.094 ±.005 [2.39 ±.127]	.102 ±.015 [2.59 ±.381]	.085 ±.005 [2.16 ±.127]	.312 ±.062 [7.92 ±1.57]
ERH-25 ENH-25	.719 ±.005 [18.26 ±.127]	.781 ±.005 [19.84 ±.127]	1.062 ±.031 [26.97 ±.787]	1.938 ±.062 [49.23 ±1.57]	.550 ±.015 [13.97 ±.381]	1.080 ±.015 [27.43 ±.381]	.546 ±.015 [13.87 ±.381]	.075 ±.010 [1.90 ±.254]	.231 ±.010 [5.87 ±.254]	.172 ±.010 [4.37 ±.254]	.125 ±.005 [3.18 ±.127]	.115 ±.015 [2.92 ±.381]	.085 ±.005 [2.16 ±.127]	.438 ±.062 [11.13 ±1.57]
ERH-50 ENH-50	1.562 ±.005 [39.67 ±.127]	.844 ±.005 [21.44 ±.127]	1.968 ±.031 [49.99 ±.787]	2.781 ±.062 [70.64 ±1.57]	.630 ±.015 [16.00 ±.381]	1.140 ±.015 [28.96 ±.381]	.610 ±.015 [15.49 ±.381]	.088 ±.010 [2.24 ±.254]	.260 ±.010 [6.60 ±.254]	.196 ±.010 [4.98 ±.254]	.125 ±.005 [3.18 ±.127]	.107 ±.015 [2.72 ±.381]	.085 ±.005 [2.16 ±.127]	.438 ±.062 [11.13 ±1.57]

Note: All resistance ranges shown conform to military specifications unless otherwise indicated.

Very large resistors are manufactured from resistive ribbon, made from Ni-Cr or constantan, which is wound on tubular ceramic or steatite cylinders, as shown in Fig. 7. These resistors, which are capable of dissipating several hundred watts, are primarily used as rheostats or ballasts in motors and electrical traction systems for high-speed trains.

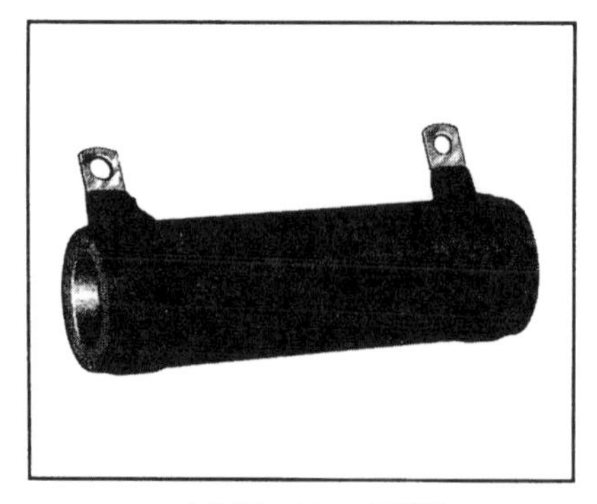

(**a**) HL, 12 to 225W

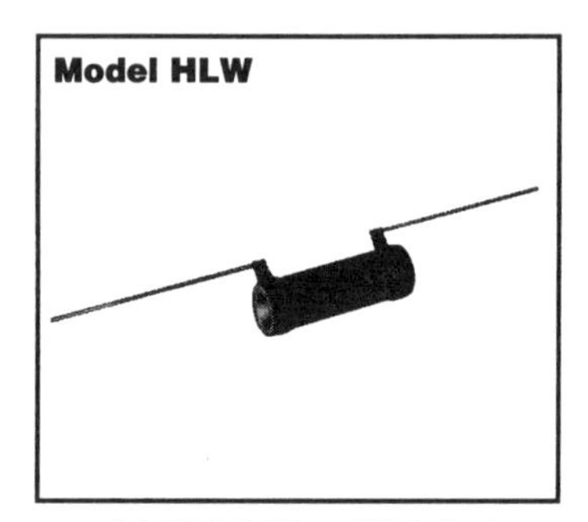

(**b**) HLM, 10 to 20W (flat)

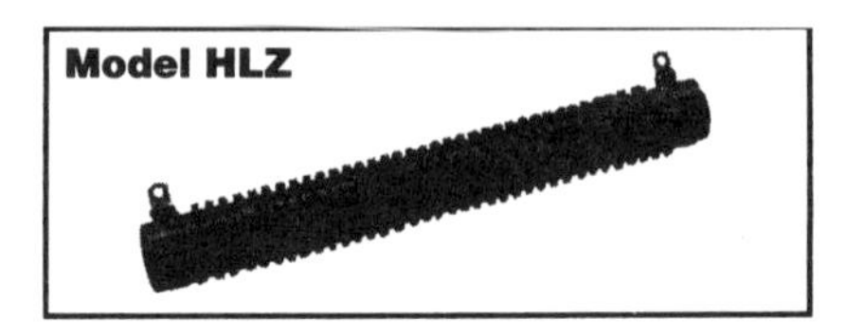

(**c**) HLW

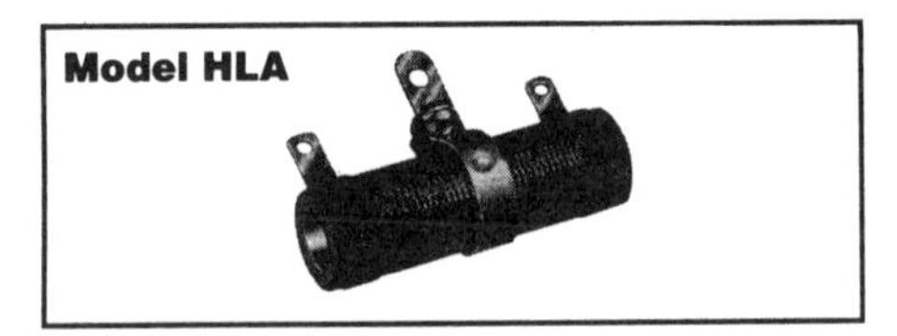

(**d**) HLA, 12 to 225W (adjustable)

Fig. 7 High-power WW resistors.

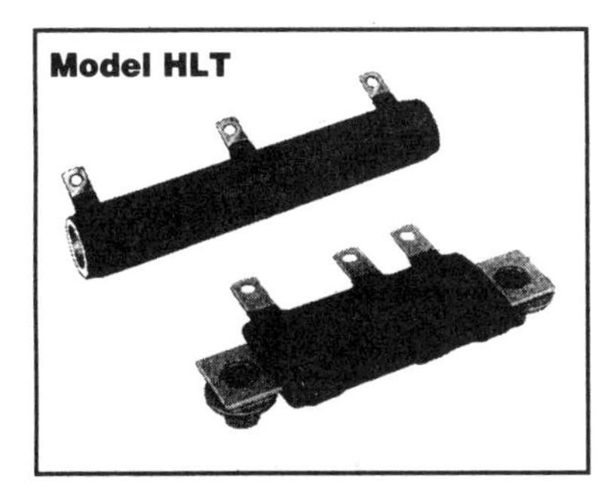

(e) HLT (tapped)

(f) HLM, 10 to 20W (flat)

Fig. 7 *(continued)*

4. LOW-OHMIC VALUE POWER RESISTORS

Measurement of high currents requires the use of shunts having low-ohmic values and high-performance electrical characteristics in terms of TCR and charge stability. Given the difficulty of manufacturing low-ohmic value wirewound resistors, manufacturers have resorted to the use of Ni-Cr or Cu-Ni resistive foil materials in place of resistive wire.

A 0.2 to 1 mm thick strip of Ni-Cr alloy foil is cut into a rectangle. After welding the copper terminal—wire or strip—a laser is used to cut a "Y" in the foil to help adjust the resistance (this includes the lead wire, whose resistance cannot be neglected) to the required value. Using this technology, manufacturers can produce resistors with ohmic values between 0.001 and 1Ω with tolerances between 1% and 10% of nominal value,[2] depending on the application. The other electrical characteristics, such as the TCR and charge deviation, are those of the foil, which, after suitable heat treatment, can reach values of 10 ppm/°C and 1000 ppm ΔR/R drift at 125°C

[2]Dale LVR series resistors, for example.

and 10,000 hours of rated load.[3] An additional advantage of this technology is the ease with which surface-mount components can be produced. Figure 8 illustrates the method of constructing a component with such electrical characteristics.

(a) Resistive element style LVR (axial).

(b) Chip resistor, style WSL (surface-mounted).

Fig. 8 Low-value Dale resistors made with a strip of metal.

[3]It is not possible to fully compensate for the influence of the lead wire on the TCR and resistance. The lead wires, having highly positive TCR values (4000 ppm/°C and higher), cannot be used to obtain an overall TCR close to 0 ppm/°C for very low ohmic values. These values can only be obtained through the use of specialized mounting technologies, which we will examine in the chapter on high-precision resistors.

(**c**) Molded surface-mounted low-value resistor, Dale model WSR.

Fig. 8 (*continued*)

5. HIGH POWER SURFACE-MOUNT RESISTORS

As discussed in the preceding sections, conventional power resistors constructed from wirewound around a ceramic tube are bulky and expensive. Cooling them also requires bulky materials and users want smaller components with the same dissipation. This is especially true for applications in portable small devices as well as in trains powered by electrical engines, where weight reduction (several tons) and size are important considerations.

To overcome these difficulties the industry has recently developed surface-mount power resistors, which provide improved dissipation along with lighter weight and smaller size. In addition to improved response times compared with wound technology, this new family of resistors enjoys the advantage of better reliability in steady-state and pulsed applications. These components are sold in the form of chips a few centimeters on a side and one or two centimeters thick. Screwed or soldered terminal connectors are located on the upper face of the component. The resistive element consists of a cermet paste that is silk-screened onto an alumina or aluminum nitride substrate to improve heat dissipation. However, this technology, when noise, load-life stability and TCR are considered, is not as good as foil technology.

We saw earlier that the critical variable controlling resistor reliability is the maximum temperature attained by the resistive element (*junction temperature*). This temperature is responsible not only for the mechanisms that lead to component degradation but for the thermomechanical constraints that affect its reliability. Simulation (see Tables 5 and 6) shows that, for a given geometry, the adhesive interfaces between ceramic, metal, and polymer play a considerable role in this. Figure 9 is a cut-away drawing of the simulated device.

The dissipation temperature is a function, not only of the constituent elements in the component (like the copper element 5 on Fig. 9) but of the quality of the interfaces (adhesives, contacts, etc.) between these different elements. Tables 5 and 6 illustrate these features for a $50 \times 60 \times 10$ mm resistor in which ΔT_{max} represents the maximum temperature difference during operation between the resistive film and aluminum base (marked 1 and 4 on Fig. 9). For a modified design, the copper element 5 shaped to contact also the aluminum base 4, the simulation showed reduced temperature rise (see Table 6). The good and poor thermal interfaces were simulated by thermal resistances of 10 and 5 $K \cdot Cm^2/W$, respectively.

We know that the limiting temperature of a cermet paste shouldn't exceed approximately 150°C. If we make use of the best possible geometry (with a heat sink), the component with poor interfaces will only be able to provide an overall dissipation of 250W at 100°C ambient temperature as opposed to 500W using a component with "good" interfaces.

As we have shown, performance is more a function of the high quality of the interface than the overall shape of the resistor. Figure 10 represents the

TABLE 5 Thermal Simulation Results

Power (W)	250	500	750	1000
ΔT_{max} good thermal interfaces	36	72	108	144
T_{max} poor interfaces	87	174	261	348

TABLE 6 Modified Heat Conducting Element

ΔT_{max} optimal	21	42	63	84
T_{max} poor interfaces	57	114	171	228

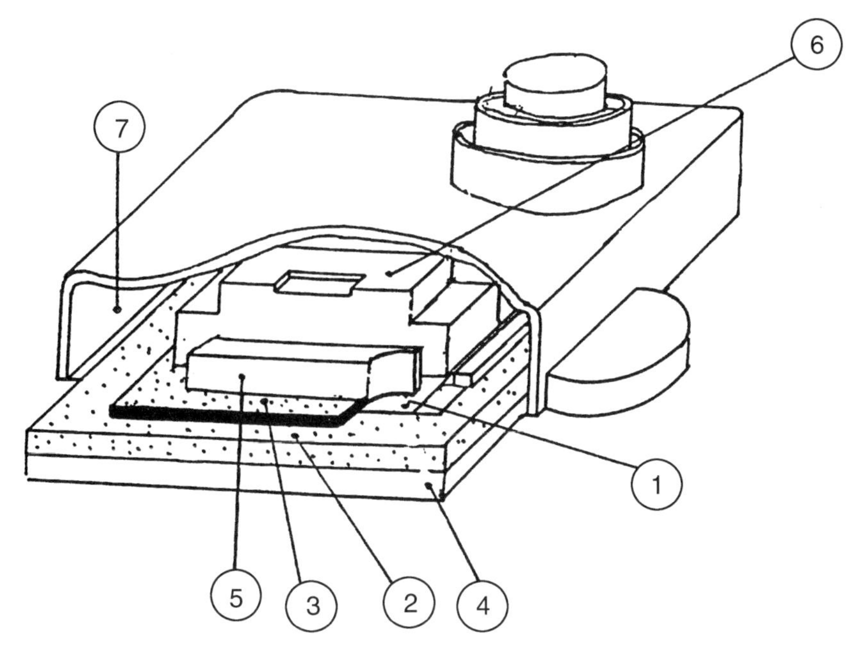

1. Resistive film
2. Alumina substrate
3. Alumina insulator
4. Aluminum base
5. Conductive element (copper)
6. Protective cover
7. Housing (resin)

Fig. 9 High-power resistor, surface-mounted (Sfernice).

results of three-dimensional heat simulations on these components. It refers to 1/4 of the device—a simplification made possible by symmetry.

6. MEASUREMENT OF HEAT DISSIPATION DURING OPERATION

When a cylindrical axial-lead resistor, as shown in Fig. 1a, is used, it is possible to measure transient and continuous heat dissipation phenomena, providing the following assumptions are made:

- We assume that the length of the cylinder is greater than its radius and that heat conduction by the wire lead is negligible compared to the heat conduction from convection.

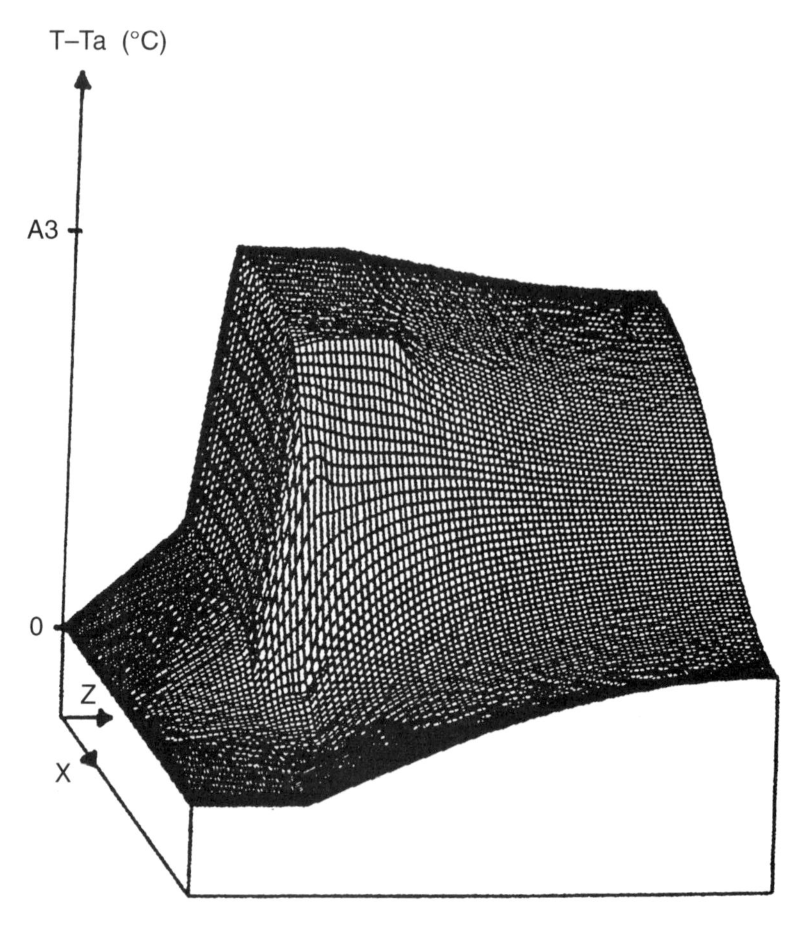

Fig. 10 Surface temperature simulation for a quarter of resistor shown in Fig. 9.

- The ceramic mass is at least one or two orders of magnitude greater than the mass of the resistive material.

Under these conditions, if ΔT is the temperature difference between the resistive layer and the outside temperature, and if we assume that the value of ΔT is at a good approximation the same in the entire volume of the component, a mathematical formula can be derived for the estimation of ΔT, such that:

$$\Delta T = \frac{I^2 R_0}{K_T - \alpha_{\mathrm{TCR}} I^2 R_0} \cdot (1 - \exp(-t/\tau))$$

where K_T is the *total* heat transfer coefficient with the exterior, in W/°C. This coefficient will take into account *all transfer mechanisms* described in Chap. 5 (plus transfer through the connections); I is the current, R_0 the value of resistance at $\Delta T = 0$, α_{TCR} the thermal coefficient of temperature, and $\tau = \rho C_p V/(K_T - \alpha_{\text{TCR}} I^2 R_0)$, and ρ, C_p, V being, respectively, the average density, the average specific heat coefficient of the component (in SI units), and its total volume.

For $t < 10^{-1} - 10^{-3}$, this equation can be approximated by: $\Delta T \approx (I^2 R_0/\rho C_p V) \cdot t$.

When t tends to infinity, the component is in thermal equilibrium with the environment and:

$$\Delta T_{\text{max}} = \frac{I^2 R_0}{K_T - \alpha_{\text{TCR}} I^2 R_0}$$

We see that when the denominator is equal to zero, $\Delta T \to \infty$ and the resistor acts as a fuse. This relation will be used for measuring K_T.

The product $\rho C_p V$ can also be estimated by means of a pulse generator, which feeds the resistor with current pulses of intensity I and various durations (10^{-2} to 10^{-1} sec). Because the TCR is known, measurements of R provide the value of junction temperature, and the linear approximation can be used to estimate $\rho C_p V$. In practice, and in order to measure these constants with any degree of accuracy, it is necessary to replace the resistive elements, wire, film, thin film or cermets with a resistive element having a TCR as high as possible. Nickel wires, oxides, etc. are well suited for this purpose.

Again, these relations are suitable for geometrically simple components. For complex form components, or great heterogeneity, other means, such as finite elements, should be used, as was the case for components described in the preceding section.

7. CONCLUSION

Despite the relative age of this technology (the first power resistors manufactured were wirewound resistors), this type of resistive component is still used for applications involving power electronics. However, to satisfy more stringent mounting requirements, the majority of them have been modified for surface-mount technology.

10

Precision Resistors and Resistor Networks

1. INTRODUCTION

The introduction of ever more sophisticated electronic measurement and data transmission systems in the 1960s for use in industrial, military, communications, and scientific applications has resulted in the increasing use of precision resistors. The performance of these components is generally two or three orders of magnitude greater than that of conventional resistors. For example, proper operation of a 20-bit analog-digital converter requires the use of resistors with ohmic values between 1,000 and 100,000Ω, tolerances of approximately 0.005% between resistors, TCR of ±2 ppm/°C, and operating drift of less than 150 ppm at 125°C and 10,000 hours of operation.

It is easy to see how the development and fabrication of such components introduced requirements that were quite different from those for resistors used in less critical applications. More stringent requirements led to the development of new resistor technologies for wirewound resistors, resistive film, foil materials deposited or glued on rigid substrates, and thin films deposited on substrates and etched to create discrete resistances or resistor networks. In the current chapter we will examine these various technologies and discuss their relative merits.

2. PRECISION WIREWOUND RESISTORS

The technology of precision wirewound resistors is not fundamentally different from that of power resistors. The only difference lies in the nature (the use of Karma alloys, for example. See Chap. 2) and heat treatment of the resistive wire prior to use—designed to improve the TCR of the alloy to about ±5 ppm/°C—and the care exerted in the construction of the resistor and its winding to minimize mechanical strain. Ohmic values are between 100 ohms and several megohms, with TCR of ±5 ppm°C. Irreversible ohmic drift at 125°C is approximately 500–700 ppm at 10,000 hours of operation. Unfortunately the winding introduces a self-induction that prevents the use of these resistors in high-frequency electronic systems, limiting them to measuring phenomena that vary slowly (frequencies below 1–2 MHz). One method of fabrication with double[1] strands of wire is currently in use and reduces self-inductance in these devices. Unfortunately this results in an improvement of no more than roughly an order of magnitude. In addition, the technology required to integrate several discrete resistors in resistive networks is extremely difficult and expensive. Figure 1 illustrates the method of fabrication and principal characteristics of some typical commercial wirewound precision resistors.

3. FOIL RESISTORS

In 1962 a small American company called Vishay Intertechnology was assigned a series of patents[2] for a new technology for manufacturing state-of-the-art precision resistors. These resistors based on strain gage technology are made as follows (see Fig. 2).

A foil of resistive Ni-Cr, or similar alloy, with a thickness of 2–8 μm is glued to a solid ceramic or glass substrate after being heat treated to adjust its temperature coefficient of resistance (TCR) to an appropriate value. A photosensitive resin is then deposited on the foil using microelectronic processes, similar to integrated circuit process technology. The photosensitive resin is exposed (photolithography) through a photographic mask representing the design of resistance circuit, which resembles a series of looping filaments. The nonexposed areas are washed off, leaving the exposed areas intact, forming a meander pattern (photoresist mask) on top of the foil. The foil areas not protected by the photoresist mask are then

[1]Vishay (Ultronix) and RCL are the principal manufacturers of these resistors.

[2]Filed by Felix Zandman *et al.*

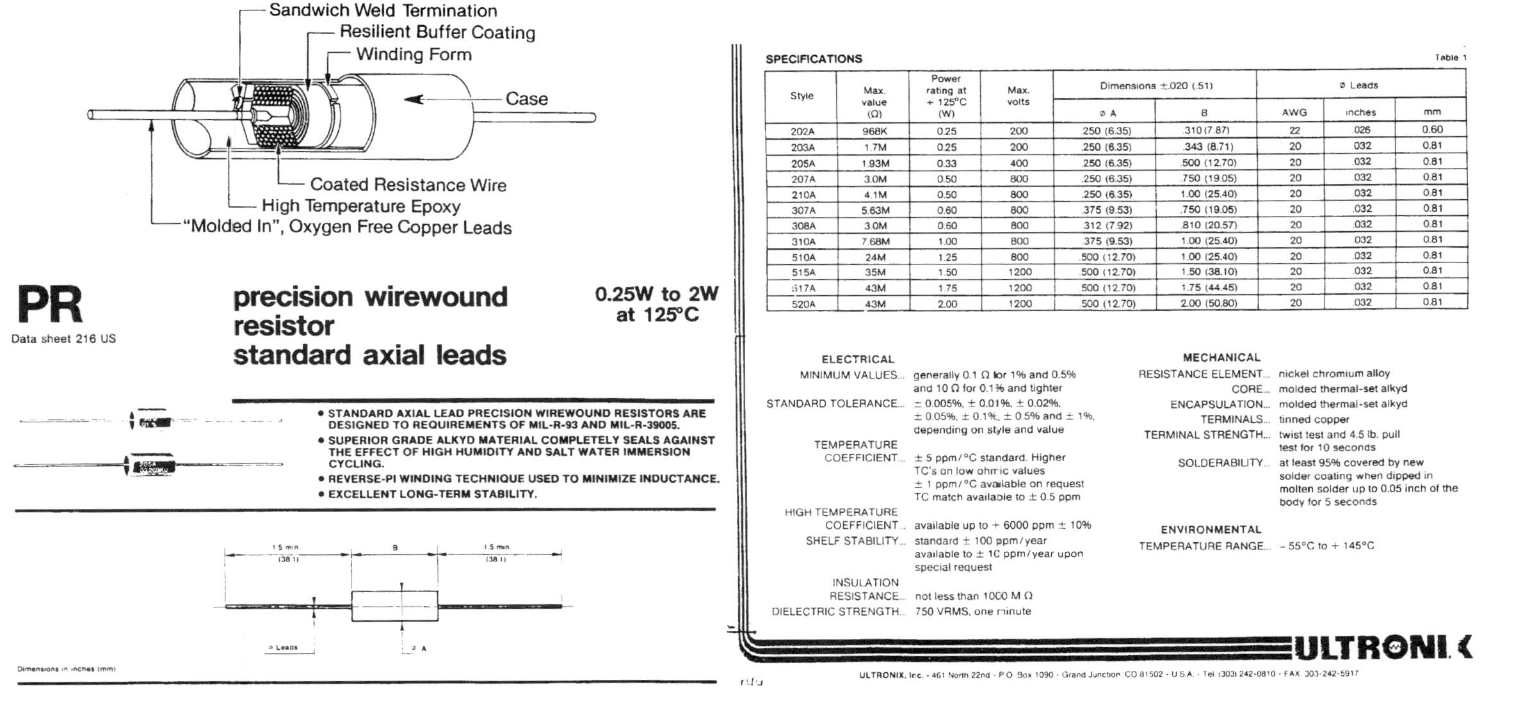

PR

Data sheet 216 US

precision wirewound resistor standard axial leads

0.25W to 2W at 125°C

- **STANDARD AXIAL LEAD PRECISION WIREWOUND RESISTORS ARE DESIGNED TO REQUIREMENTS OF MIL-R-93 AND MIL-R-39005.**
- **SUPERIOR GRADE ALKYD MATERIAL COMPLETELY SEALS AGAINST THE EFFECT OF HIGH HUMIDITY AND SALT WATER IMMERSION CYCLING.**
- **REVERSE-PI WINDING TECHNIQUE USED TO MINIMIZE INDUCTANCE.**
- **EXCELLENT LONG-TERM STABILITY.**

SPECIFICATIONS — Table 1

Style	Max. value (Ω)	Power rating at + 125°C (W)	Max. volts	Dimensions ±.020 (.51)		ø Leads		
				ø A	B	AWG	inches	mm
202A	968K	0.25	200	.250 (6.35)	.310 (7.87)	22	.026	0.60
203A	1.7M	0.25	200	.250 (6.35)	.343 (8.71)	20	.032	0.81
205A	1.93M	0.33	400	.250 (6.35)	.500 (12.70)	20	.032	0.81
207A	3.0M	0.50	800	.250 (6.35)	.750 (19.05)	20	.032	0.81
210A	4.1M	0.50	800	.250 (6.35)	1.00 (25.40)	20	.032	0.81
307A	5.63M	0.60	800	.375 (9.53)	.750 (19.05)	20	.032	0.81
308A	3.0M	0.60	800	.312 (7.92)	.810 (20.57)	20	.032	0.81
310A	7.68M	1.00	800	.375 (9.53)	1.00 (25.40)	20	.032	0.81
510A	24M	1.25	800	.500 (12.70)	1.00 (25.40)	20	.032	0.81
515A	35M	1.50	1200	.500 (12.70)	1.50 (38.10)	20	.032	0.81
517A	43M	1.75	1200	.500 (12.70)	1.75 (44.45)	20	.032	0.81
520A	43M	2.00	1200	.500 (12.70)	2.00 (50.80)	20	.032	0.81

ELECTRICAL

MINIMUM VALUES... generally 0.1 Ω for 1% and 0.5% and 10 Ω for 0.1% and tighter

STANDARD TOLERANCE... ± 0.005%, ± 0.01%, ± 0.02%, ± 0.05%, ± 0.1%, ± 0.5% and ± 1%, depending on style and value

TEMPERATURE COEFFICIENT... ± 5 ppm/°C standard. Higher TC's on low ohmic values ± 1 ppm/°C available on request TC match available to ± 0.5 ppm

HIGH TEMPERATURE COEFFICIENT... available up to + 6000 ppm ± 10%

SHELF STABILITY... standard ± 100 ppm/year available to ± 10 ppm/year upon special request

INSULATION RESISTANCE... not less than 1000 M Ω

DIELECTRIC STRENGTH... 750 VRMS, one minute

MECHANICAL

RESISTANCE ELEMENT... nickel chromium alloy

CORE... molded thermal-set alkyd

ENCAPSULATION... molded thermal-set alkyd

TERMINALS... tinned copper

TERMINAL STRENGTH... twist test and 4.5 lb. pull test for 10 seconds

SOLDERABILITY... at least 95% covered by new solder coating when dipped in molten solder up to 0.05 inch of the body for 5 seconds

ENVIRONMENTAL

TEMPERATURE RANGE... – 55°C to + 145°C

ULTRONIX

ULTRONIX, Inc. - 461 North 22nd - P.O. Box 1090 - Grand Junction, CO 81502 - U.S.A. - Tel. (303) 242-0810 - FAX 303-242-5917

Fig. 1 Construction and characteristics of precision wirewound resistors.

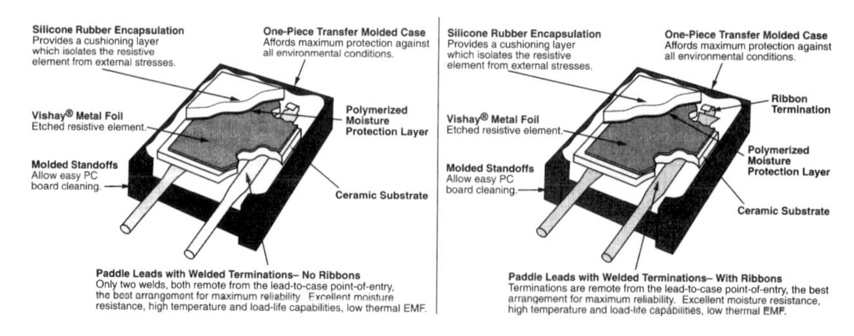

Fig. 2 Construction of a foil resistor.

etched by means of electrolytic or chemical processes, reproducing the design of the mask. This step creates hundreds of resistive filaments in series or parallel such that the resistance of the circuit reaches the desired value. Such resistance elements are usually produced in an array of several hundred on a wafer. The next process step is to singulate resistor chips from the wafer.

> It is important that the chip—resistive alloy/cement/substrate—be free from outside mechanical stress. The TCR principle is as follows: The resistance of the foil in its free state (before cementing to the ceramic substrate) *increases* with temperature. Since the foil is cemented to the ceramic substrate and since the coefficient of thermal expansion of ceramic is smaller than the coefficient of thermal expansion of the foil, as the temperature of the foil rises, the foil undergoes compressive stress and therefore the resistance *decreases.* If both phenomena are equal and of opposite sign the resulting change is close to zero. This is the essence of the low TCR of foil resistors. Section 3.1 below describes in detail the mechanism of TCR compensation.

To complete the resistor, leads are soldered or welded to each end of the resistive circuit on the chip. The foil resistors are then usually calibrated (see Fig. 3) in discrete increments (unlike thick-film and thin-film resistors whose values are adjusted by continuous laser abrasion). The foil method consists of opening designed-in numeric value shorting bars resulting in known incremental adjustment to final value and tolerance. Their adjustment to a given final value of resistance R is made according to the following procedure. Trimming points $i = 1, \dots p$, where p is the total number of trimming points, are designed according to a method based on geometrical series. In order to illustrate the method, we may use a geometrical series with a ratio $1/2$. Then a cut in the trimming point i increases the resistance by $R_0/2^i$, where R_0 is the value of resistor at the start of the trimming operation. If we assume that $p \to \infty$ (in that case, the sum value

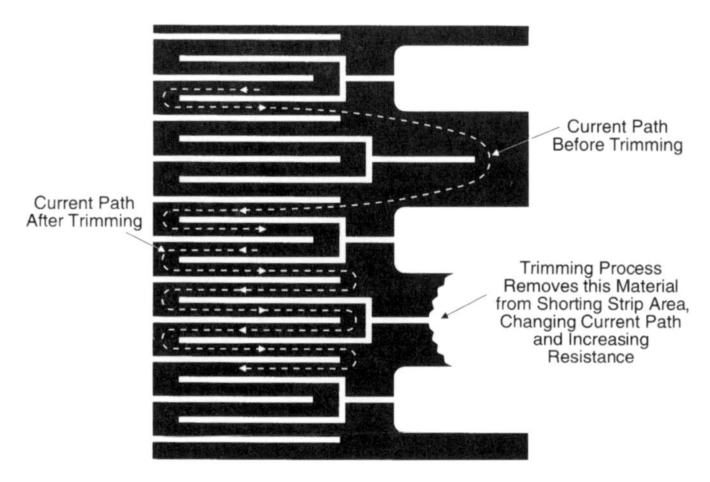

Fig. 3 Grid of a foil resistor.

of the series is 2, to be compared with the multiplier coefficient of Vishay's S102C, which is equal to 1.35–1.5) then:

$$\frac{R - R_0}{R_0} \sum_{i=1}^{\infty} \frac{\alpha_i}{2^i} \qquad \alpha_i \in [0,1]$$

Since in a real chip p is finite, the adjusted resistance R_S closest in value to R, the target nominal value which has to be obtained, is given by:

$$R_S = \left(1 + \sum_{i=1}^{p} \frac{\alpha_i}{2^i}\right) \cdot R_0$$

Obviously, $\alpha_i = 1$ if the trimming point is cut; 0 if it is not. The choice of trimming points is then determined by the following algorithm:

$$R_i = R_0$$

For $i = 1$ to $i = p$ start

if
$$\frac{R - R_i}{R_0} \geq \frac{\alpha_i}{2^i}$$

then
$$\alpha_i = 1$$

$$\alpha_i = 1 \text{ (the line must be trimmed)}$$

otherwise

$$\alpha_i = 0 \text{ (the line is not trimmed)}$$

Stop

To obtain the requested value, lines are cut successively and the resistance is measured after each cut. The algorithm determines the next trimming point accordingly, until the requested values of resistance and tolerance are obtained.

In the foil technology, this choice of trimming method is necessary. In effect, the smallest thickness that can be achieved in the rolling of Ni-Cr foils is about 2.5 micrometers. As a result, the resistance is about $0.6\Omega/\square$, a value very small by comparison with other technologies such as thin or thick films. When tight tolerances (0.01% typical; 0.001% achievable) on the nominal values must be obtained, the classic methods of trimming used in thin or thick films cannot be used. However, the management of real estate occupation and homogeneity of power density all over the chip becomes more difficult than with these alternative technologies.

The smaller trimming weights for these tight tolerances are obtained by the implementation of several series-parallel elements of resistance and necessitates a larger real estate than the others. Under these conditions, finding the optimal compromise between these somewhat contradictory factors (surface vs. precision) necessitates the use of rather powerful computational tools to design these components and define the best trimming algorithm.

The calibrated resistor is next coated with various protective layers to reduce effects of humidity and mechanical stress (from molding or user handling). The resistors are then encapsulated or molded, marked, and are now complete.

3.1 Physical Principles of TCR Control and Adjustment

It is worthwhile to develop a detailed mechanism for controlling the *TCR*. We saw in Chap. 4 that the TCR of an alloy depends not only on the physical properties of conductivity associated with the structure of the material, but also through Grüneisen's equation and the Bridgman constant, the mechanical stress on the crystal lattice itself (see Chap. 2).

$$\frac{dR}{R} = \frac{d\rho}{\rho dT} \cdot \Delta T + 2 \cdot \frac{dL}{L} + (C - 1) \cdot \frac{dV}{V} \tag{1.1}$$

Here, as in Chap. 2, ρ is the resistivity of the material, C the Bridgman constant, ΔT the difference in temperature between the ambient and the effective temperature of the material, and V the volume of the material. The term $\partial\rho/\rho dT$ is the coefficient of temperature of the material at constant volume, which we designate by TC_ρ.

We assume that the resistive material in its free state (a foil before being cemented to the substrate, or a free wire) can expand freely. Under these conditions, if α_R is its coefficient of expansion, L its length, l its width, and t its thickness, the resistance is equal to $R = \rho\, L/lt$. When the temperature difference is given by ΔT, we have $dV/V = dL/L + dl/l + dt/t = 3\alpha_R\Delta T$ and $dL/L = \alpha_R\Delta T$. By introducing these values into Eq. 1.1, we have:

$$\frac{dR}{R} = TC_\rho \cdot \Delta T + (2 + 3(C - 1))\alpha_R\Delta T \tag{1.2}$$

from which we derive

$$TCR = \frac{dR}{R \cdot \Delta T} = TC_\rho + (2 + 3(C - 1))\alpha_R \tag{1.3}$$

(For materials in which $C \cong 1$, the TCR is independent of changes in volume.)

The coefficient of expansion of metals and metallic alloys is approximately $10 \cdot 10^{-6}/°K$. If the foil is made of a pure metal ($TC_\rho \cong 4000$ ppm/°K), the term in α_R has little effect on the outcome.

In Ni-Cr metal alloys, however, in which TC_ρ is only a few ppm/°K, changes in dimension as a function of temperature or other parameters play a significant role in the variation in resistivity. Precision wirewound resistors, for example, are constructed in such a way that the wire can expand freely, and Eq. 1.3 then applies.

In the case of foil resistors, however, we have to account for the effect of mechanical stress introduced by the substrate. In Fig. 2 the resistive alloy foil is glued to the substrate. To calculate the effect of adhesion on the TCR of the resistor, we have to make the following two assumptions:

- The foil is thin compared to the thickness of the substrate, which is assumed to be isotropic and causes the deformations, dL/L and dl/l (in practice, the substrate material is glass or polycrystalline alumina ceramic, between 0.3 and 0.8 mm thick. The foil is between 2 and 10 μm thick.)

- The upper surface of the foil is free from stress ($\sigma_t = 0$). We can therefore neglect the stress from a thin protective layer.[3]

Under these conditions we can apply Hooke's law,[4] which expresses the relationship between stress (σ_L, σ_l) and strain in continuous media through the use of a constant of elasticity associated with the materials. Here, we can write Hooke's law as follows:

$$\frac{\Delta L}{L} = \frac{1}{E}(\sigma_L - \mu_R \sigma_1) + \alpha_R \Delta T = \alpha_S \Delta T \tag{1.4}$$

$$\frac{\Delta l}{l} = \frac{1}{E}(\sigma_L - \mu_R \sigma_l) + \alpha_R \Delta T = \alpha_S \Delta T \tag{1.5}$$

$$\frac{\Delta t}{t} = \frac{1}{E}[-\mu_R(\sigma_L + \sigma_l)] + \alpha_R \Delta T \tag{1.6}$$

E is Young's modulus and μ_R Poisson's ratio for the foil, α_R its coefficient of expansion, and α_S the coefficient of expansion of the substrate. ΔT is the temperature change of the unit as a whole.[5] The sum of Eq. 1.4 and 1.5 yields:

$$\sigma_L + \sigma_l = \frac{E}{1 - \mu_R} \cdot 2(\alpha_S - \alpha_R) \cdot \Delta T \tag{1.7}$$

By inserting this in Eq. 1.6, we can calculate $\Delta t/t$.

$$\frac{\Delta t}{t} = -\frac{\mu_R}{1 - \mu_R} \cdot 2(\alpha_S - \alpha_R) \cdot \Delta T + \alpha_R \Delta T \tag{1.8}$$

Because we assume that $\Delta L/L = \Delta l/l = \alpha_S \cdot \Delta T$, we can write:

$$\frac{\Delta V}{V} = \frac{\Delta L}{L} + \frac{\Delta l}{l} + \frac{\Delta t}{t} = \left(\frac{1 - 2\mu_R}{1 - \mu_R} 2\alpha_S + \frac{1 + \mu_R}{1 - \mu_R} \cdot \alpha_R\right) \cdot \Delta T \tag{1.9}$$

[3]Additional detail can be found in papers about effects of protective coatings by F. Zandman and Post and about TCR in foil resistors by F. Zandman and Stein, see App. I.

[4]For additional information interested readers may refer to the literature on the mechanical theory of continuous media, such as Dugdale, D. S. and Ruiz, C. *Elasticity for Engineers.* London: McGraw-Hill, 1971.

[5]These equations are valid only for reversible elastic phenomena, in which strain effects have not yet become permanent.

We can now write $\Delta R/R$ and TCR as follows:

$$\frac{\Delta R}{R} = TC_{\rho} \cdot \Delta T + 2\alpha_S \cdot \Delta T + (C-1)\left(\frac{1-2\mu_R}{1-\mu_R} \cdot 2\alpha_S + \frac{1+\mu_R}{1-\mu_R} \cdot \alpha_R\right) \cdot \Delta T \tag{1.10}$$

$$TCR = TC_{\rho} + \left(1 + (C-1) \cdot \frac{1-2\mu_R}{1-\mu_R}\right) \cdot 2\alpha_S + (C-1) \cdot \frac{1+\mu_R}{1-\mu_R} \cdot \alpha_R \tag{1.11}$$

By subtracting the TCR of the unstressed foil, given in Eq. 1.3, from this equation, we can express the effect of the substrate-foil bond of the TCR:

$$\Delta TCR = 2 \cdot (\alpha_S - \alpha_R) + (C-1) \cdot \frac{1-2\mu}{1-\mu} \cdot 2(\alpha_S - \alpha_R) \tag{1.12}$$

which can be written as

$$\Delta TCR = 2 \cdot (\alpha_S - \alpha_R)\left[1 + (C-1) \cdot \frac{1-2\mu_R}{1-\mu_R}\right] \tag{1.13}$$

Equation 1.13 shows that if the substrate expands along with the foil ($\alpha_S = \alpha_R$), there is no stress and the above equation reduces to Eq. 1.3[6] for the value of TCR given in Eq. 1.11.

If $C \cong 1$, changes in volume do not affect TCR. This is true for Ni-Cr, which is used in strain gages, where the gage factor is equal to 2. As a result, changes in TCR in precision foil resistors can be expressed as $\Delta TCR \cong 2\Delta\alpha$, which can be shown through experiment ($\Delta TCR \cong -13$ppm/°C when the

[6]When the stress is mechanical, we can use the same method to determine the gage factor. Under these circumstances, we can show that when a strain field, ε_x, ε_y, and γ (the strain of the right angle associated with shear), acts on the resistor, the resistance change can be expressed by the following equation:

$$\frac{\Delta R}{R} = 2\varepsilon_X + 2g(\theta) + (C-1)\frac{1-2\mu_R}{1-\mu_R} \cdot (\varepsilon_X + \varepsilon_Y)$$

$$g(\theta) = \frac{\varepsilon_X + \varepsilon_Y}{2} \cdot (\cos(2\theta) - 1) + \frac{\gamma}{2} \cdot \sin(2\theta)$$

Ni-Cr alloy is glued to an alumina substrate). Obviously, this value will differ for other metals and alloys, and different substrates, since it depends on the difference between the various coefficients of expansion.

The equation for ΔTCR is valid if we assume that the bonding of the foil to the substrate is such that any strain in the substrate is completely transmitted to the foil. In practice, given the experimental conditions under which bonding takes place, our assumption is not entirely correct. The measured temperature coefficient is basically a combination of nonlinear elements, which can be calculated and are generally known by the resistor manufacturer. For the most part, they are the sum of the applied first- and second-order mechanical or thermomechanical stresses. Nevertheless, the equation is helpful in understanding the physical phenomena involved and can be used to determine their orders of magnitude.

Of course the nature of the resistive foil affects all the resistor's characteristics, but other elements involved in its construction come into play. Load-induced drifts, for example, are very sensitive to the quality of the bonding materials used to attach the foil to the substrate. The resins used, which are epoxy-based, are specially prepared so that their Martens[7] point is as high as possible. This helps avoid the introduction of deviations in the mechanical stresses arising from the action of the substrate on the foil. The effect of the other materials on the foil must also be carefully estimated. These include the protective resins, leads, solder, etc.

When the resistor is operating in a pulsed mode, as is true of an A/D or D/A converter, the adiabatic heating of the resistive elements during the few microseconds following the establishment of the voltage step results in a brief negative drift of the resistance, which can adversely affect the operation of the circuit to which it is connected. Appendix I covers this problem in greater detail, and lists orders of magnitude of the change and some of the methods used to prevent it.

The longitudinal gage factor K, which is measured when ε acts in the longitudinal direction of the resistance, $\theta = 0$, is then equal to $K = 2 + B(1 + \mu)$. The constant B depends on the physical characteristics of the material and is equal to $B = (C - 1) \cdot (1 - 2\mu_R)/(1 - \mu_R)$. For materials commonly used in strain gages, $C \cong 1$ and $K \cong 2$. The drift of the gage, regardless of the bonding orientation θ on a substrate deformed by $\varepsilon_x = \varepsilon$, $\varepsilon_y = -\mu\varepsilon$ and γ is given by: $\Delta R/R = [\mathrm{K} + (1 + \mu)(\cos(2\theta) - 1]\varepsilon + \gamma \cdot \sin(2\theta)$, where μ is Poisson's ratio for the substrate. Readers interested in the theory and practice of strain gages should consult the *Encyclopédie Vishay d'analyse des contraintes,* distributed by Vishay Inc., Malvern, PA or the literature of Measurements Group, a division of Vishay, P.O. Box 27777, Raleigh, NC 27611, internet address: www.measurementsgroup.com.

[7]The softening point of the material, generally a polymer, used for bonding.

FEATURES

- VH102K Series
  Maximum Temperature Coefficient of Resistance:
  ±2.5 ppm/°C (0°C to +25°C and +25°C to +60°C)
  ±2.5 ppm/°C (–55°C to +25°C and +25°C to +125°C)
- VH102 Series
  Nominal Temperature Coefficient of Resistance:
  +0.6 ppm/°C (0°C to +25°C); –0.6 ppm/°C (+25°C to +60°C);
  +2.2 ppm/°C (–55°C to +25°C); –1.8 ppm/°C (+25°C to +125°C)
- VHS555 Temperature Coefficient of Resistance
  ±5 ppm/°C (–55°C to +125°C);
- Selected TCR Tracking: to 0.5 ppm/°C (matched sets)
- Load-Life Stability: ±0.0025% maximum ΔR at 0.1 watt, +60°C, 2,000 hours (VHS555)
- Power Rating: 0.6 Watts at +70°C; 0.3 Watts at +125°C
- Resistance Tolerance (Initial Resistance Accuracy):
  ±0.005% tightest to ±1.0% loosest
- Resistance Range: 1 Ω to 150K Ω

Fig. 4 Features of a hermetically sealed foil resistor.

When all the conditions necessary for the correct performance of the component are fulfilled, however, its characteristics are remarkable, as shown in Fig. 4. The base resistive material consists of a special Ni-Cr alloy, which was described in Chap. 3. This same alloy is also used in precision wirewound resistors.

The geometry of the resistive circuit and the cross-sectional area of neighboring conductors (see Fig. 2) result in a high frequency response of approximately 100 MHz. Current manufacturing technology is now used to fabricate chip resistors, which can be assembled in a single package (silicon, for example) to create high-performance resistor networks. Figure 5 illustrates a typical resistor.

4. THICK-FILM PRECISION RESISTORS

The thick-film materials described in Chap. 4 are used to produce excellent precision resistors. Although these can be made to resemble wirewound or foil resistors, thick-film resistors are primarily sold in the form

Fig. 5 Precision resistor network.

of surface-mount chip resistors. Figure 6 illustrates some possible configurations of thick-film resistors and Fig. 7 shows their typical production flow.

Screen printing machines and automatic ovens are used to fabricate these resistors, resulting in high rates of production and excellent reproducibility.

(a) Thick-Film Chip Resistors

(b) Dual-In-Line, Molded

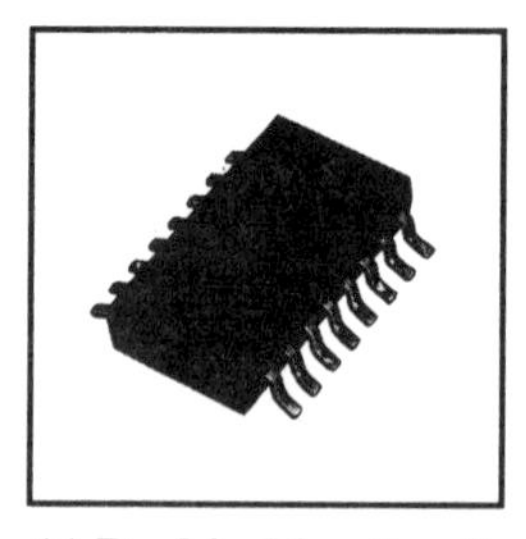

(c) Dual-in-Line, Small Outline Molded Dip

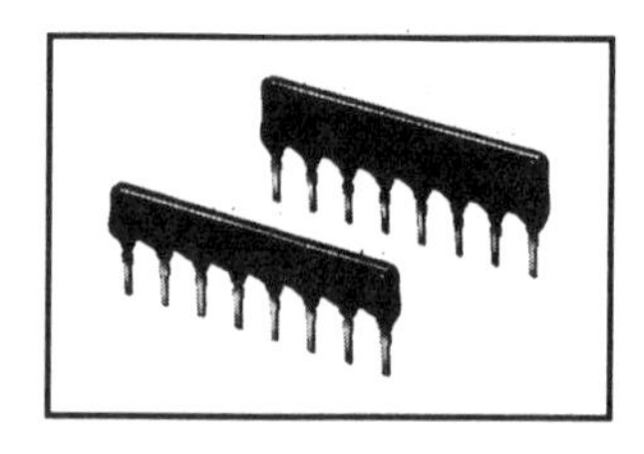

(d) Single-In-Line, Coated 4 thru 10 Pins

Fig. 6 Thick-film resistors and networks.

1) Fab. mask
2) Ink

| Conductor
| Dielectric => 3) Screen printing => Drying and Baking
| Resistor

4) Adjustment 5) | => Networks | => 6) Inspection => Packaging
| => Discrete components

Fig. 7 Production flow of thick-film resistors.

The substrate is generally a ceramic material whose physical properties are given in Table 1.

The ink (prior to firing, the thick-film, or cermet, is mixed with volatile solvents and polymers in order to get rheology values suitable for the screening process. The color of the said mixed materials is from gray to black. The name "ink" was coined for this reason) is screen printed through a silkscreen mask onto the ceramic substrate. After drying in an infrared oven, designed to evaporate the volatile solvents, the component is baked at temperatures up to 800–900°C. During the first phases of fabrication, Ag-Pd, Pt-Pd, or Au-Pd conductors are deposited. After baking, the dielectric inks (if the circuit is made in several layers or contains bridges between conductors) and resistive elements are deposited using the same process. The conductive particles are bound with a type of borosilicate glass containing Al_2O_3, PbO, Bi_2O_3, or related compounds, designed to improve its

TABLE 1 Physical Properties of Substrates

	Al_2O_3 (96.0%)	Al_2O_3 (99.5%)	AlN	SiC
Coef. Expansion ppm/°C 25–300°C	6.4	6.6	4.2	3.7
Thermal Conductivity $W \cdot m^{-1} \cdot K^{-1}$	25	34	180	150
Dielectric constant at 1 MHz and 25°C	10	10	8	40
Breakdown voltage kV/mm 25°C	18	18	22	1
Resistivity $\Omega \cdot$ cm 25°C	$>10^{14}$	$>10^{14}$	$>10^{14}$	$>10^{14}$
T° max. operational °C	1500	1600	1800	1900

mechanical and thermal properties. The most commonly used glass, which has a coefficient of expansion comparable to alumina, is made of 63% PbO, 25% B_2O_3, and 12% SiO_2. The nominal resistance value is adjusted by abrading the screened ink by means of a YAG or other laser. To complete the fabrication cycle, the substrates are broken up into individual chips and sorted. Figures 8 and 9 illustrate some of the equipment used by Dale, the principal manufacturer of thick-film network resistors. Custom resistor networks can also be manufactured using the same technology.

For the most part, the electrical performance of these devices depends on the resistive materials used and the baking operation. Starting from identical paste compositions, as described in Chap. 4, different bake parameters can be used to modify the *TCR* or stability to a significant extent. The primary factor affecting the electrical characteristics of the resistors appears to be the nature of the residual gas in the baking ovens.

The thermal and mechanical stresses on the resistive foil arising from the substrate and the environment, which was described above, also affect cermets, resulting in drift of their ohmic value and variations in the *TCR*. These effects are amplified by the fact that the gage factor of cermets is between 6 and 12. Table 2 shows that a given strain ε_x can cause significant

Fig. 8 Screen printing machine—high volume, rotary.

Fig. 9 Baking oven and firing profile.

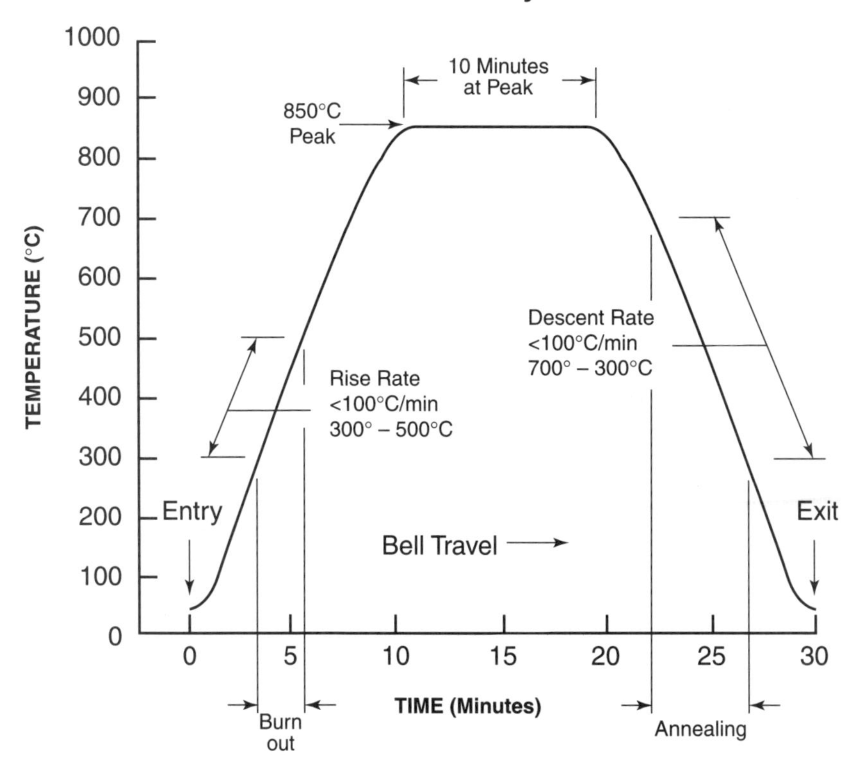

Fig. 9 (*continued*)

TABLE 2 Resistance Change Due to Mechanical Strain

Resistivity	ΔR/R(L)ppm	K(L)	ΔR/R(1)ppm	K(1)	K(L)−K(1)
Dupont 4800					
100 Ω/■	515	5.6	1036	3.7	1.9
1 KΩ/■	713	7.7	1464	5.2	2.5
10 KΩ/■	880	9.5	1835	6.5	3.0
100 KΩ/■	1066	11.6	2683	9.6	2.0
1 MΩ/■	1055	11.5	2467	8.8	2.7

drift, depending on whether it acts longitudinally or transversely to the resistors in a given network.

K(L) and K(l) represent the longitudinal and transverse gage factor. Experiments have been conducted with an alumina substrate that was subjected to a given range of values. Under these conditions ε_x is 280 ppm when the resistor is in a transverse position and 92 ppm when it is longitudinal. Under the same experimental conditions, the relative deviation between resistors made from a metallic alloy would have been smaller by an order of magnitude.

Figures 10 and 11 represent the geometry and performance that can be expected from these devices. Thick-film devices probably provide the best performance that can be obtained in resistive materials and networks in terms of reliability and price. These components are used in any electronic application requiring the use of resistors, although they are most frequently found in computer and telecommunications devices.

The performance of these components, although not as good as foil or thin-film resistors, is still remarkable. The average *TCR* falls between 50 and 200 ppm/°C for practically all ohmic values, which range between 100Ω and a few dozen MΩ, up to a maximum of 2 MΩ/mm². Operating

Fig. 10 Geometry of thick-film resistors and arrays (Dale Model CRCW).

FEATURES
- Choice of 4 or 8 terminal package with 2 or 4 isolated resistors
- Automatic placement capability
- Flow solderable
- Inner electrode protection
- Thick-film resistance element
- Wrap around termination
- Standard E-24 (± 2% and ± 5%) and E-96 (± 1%)
- Operating temperature range of - 55°C to + 150°C

ELECTRICAL SPECIFICATIONS
Resistance Range: 10 ohm to 1 Megohm.
Tolerance: ± 1%, ± 2%, ± 5%.
Temperature Coefficient: (- 55°C to + 125°C) ± 200PPM/°C standard.
Power Rating: 1/16 watt @ + 70°C per element.
Dielectric Strength: 100 VAC.
Insulation Resistance: 1000 Megohm.
Operating Temperature: - 55°C to + 150°C.
Maximum Working Voltage: 50 volts. Rated continuous working voltage (RCWV) shall be determined from RCWV = $\sqrt{\text{of Rated Power x Resistance Value}}$ or 50 volts whichever is less.

Fig. 11 Geometry and performance of thick-film resistors and arrays (Dale Model CRA065).

stability is approximately 1% or less at 125°C and 10,000 hours of operation, substantially higher than for wire, foil, or thin film. To achieve these performances, resistive inks must undergo additional thermal treatment with a mixture of appropriate gases. It should be pointed out, however, that the *l/f* noise for thick-film components is greater by two or three orders of magnitude than in wirewound foil or thin-film resistors.

5. THIN-FILM PRECISION RESISTORS

Thin-film resistors and networks are made as shown in Fig. 12.

This production flow is common to all precision chip resistors or networks. The substrates, between 3 and 6 inches, are generally 99.5% alumina with a 1–5 μm grain size. For some applications, silicon substrates are used. Under these conditions a 1–2 μm dielectric layer of SiO_2 or Si_2N_3 is deposited on the surface and functions as an insulator. Several types of resistive materials or alloys, similar to those described in Chap. 4, are deposited using different technologies such as vacuum, plasma, or ion deposition systems. Ni-Cr alloy (50:50) is very often used by the manufacturers. When a low *TRC* ($\leq$ 2 ppm/°C) for the network is a must, it is doped during deposition by O_2, N_2 molecules or other compounds. Resistive deposits of Ta_2N are also used. This material has the advantage of being self-passivating when the resistors are used in humid environments. In the opposite case (Ni-Cr), a Si_3N_4 passivation layer is deposited on the resistive material. Conductors are generally made of gold or aluminum and vacuum deposited. Figures 13 to 15 illustrate some typical equipment used to fabricate resistors and networks. The components are mass produced and 4,000–5,000

1. Deposition
   1.1. Cleaning of substrate
        chemical, or
        plasma ashing, or
        ion beam cleaning
   1.2. Deposition of the resistive layer
        sputtering, or
        ion beam deposition
   1.3. Deposition of conductive layer
        electron gun evaporation, or
        Joule's evaporation
2. Patterning (twice—for resistive and for conductive layers)
   2.1. Photoresist spinning and drying
   2.2. Exposure with UV light through mask
   2.3. Development and curing of photoresist
   2.4. Pattern etching
        plasma, or
        ion mill, or
        chemical
   2.5. Photoresist stripping
3. Heat treatment: To give the resistors the right structure which governs stability, sheet resistivity and *TCR*—an iterative process controlled by measuring the last two.
4. Glassivation
   4.1. Deposition
        Plasma (PECVD) of Silicon Nitride (Si_3N_4)
   4.2. Plasma etching of Si_3N_4 over resistor's terminations
5. Laser trimming and secondary process
   5.1. Laser trimming to precise ohmic values
   5.2 Saw dicing
   5.3. Conditioning on waffle packs (or assembly on packages)
   5.4. Visual inspection
   5.5. Interconnection by wire bonding
   5.6. Sealing of package (or molding)
   5.7. Screening, testing, and inspection

Fig. 12 Production flow of thin-film resistors.

resistors at a time can be made on a 4×4 inch wafer. Production must be carried out, however, under the same environmental conditions (clean room, deionized water, etc.) used to fabricate active semiconductor circuits. Under these conditions it is possible to produce resistive networks containing as many as 15–40 elements on a 5×5 mm^2 chip (Fig. 16). This

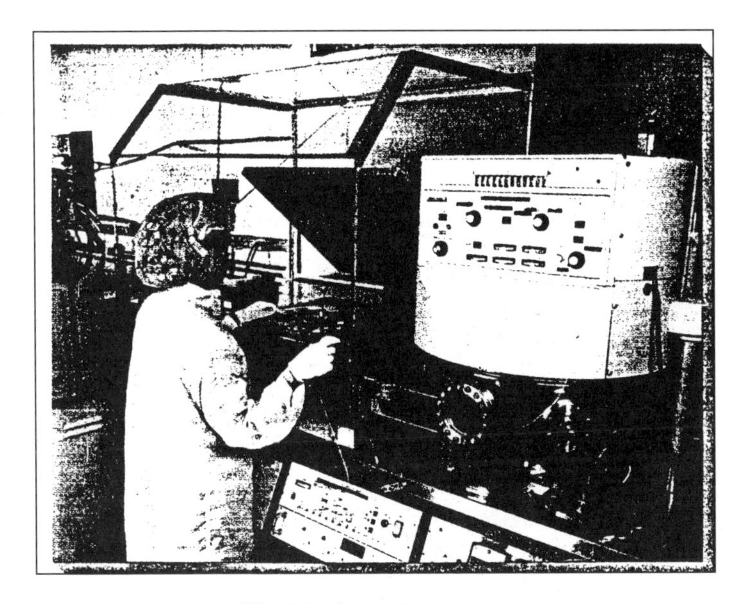

Fig. 13 Sputtering system.

Fig. 14 Laser trimmer.

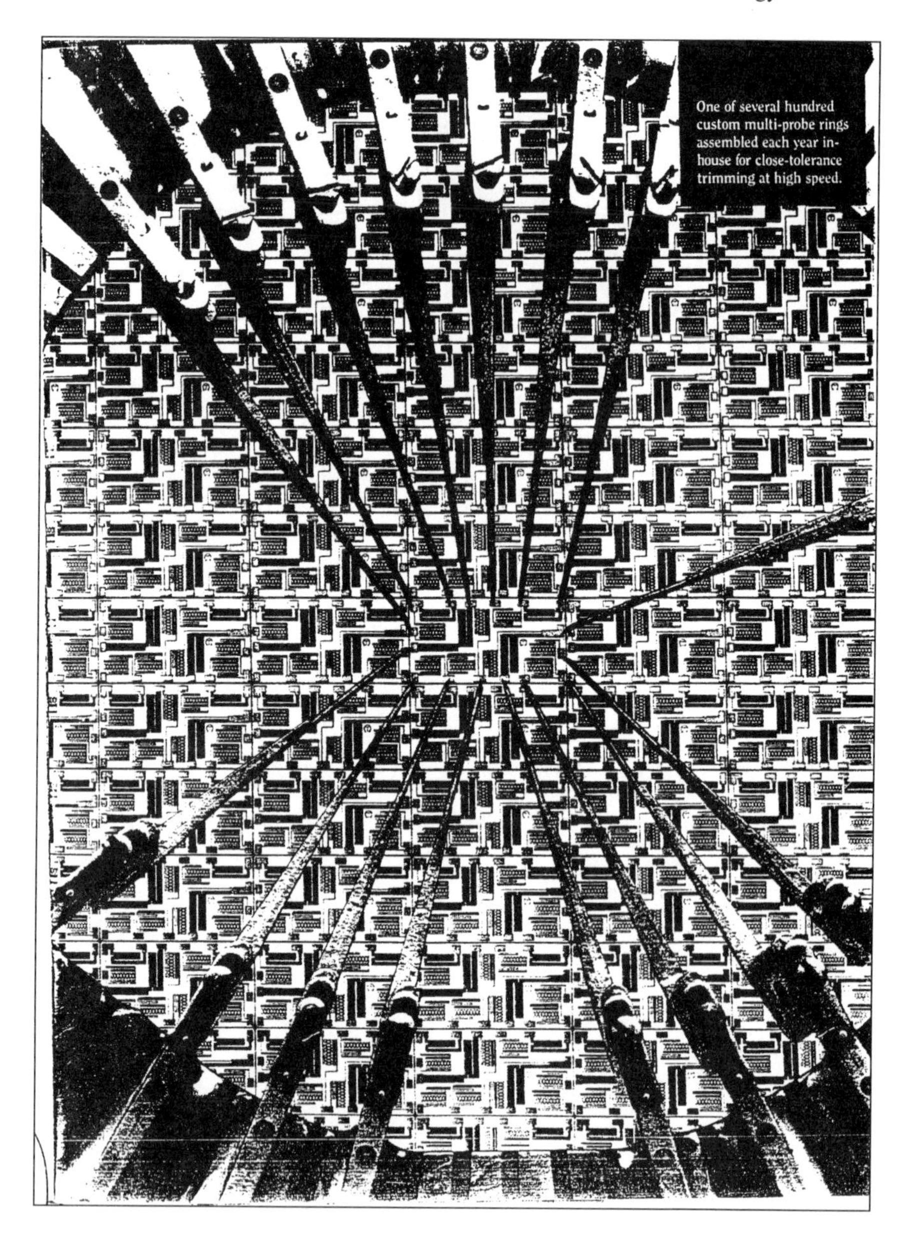

Fig. 15 Probe ring.

requires the use of plasma or ion-beam technologies to create etched components with dimensions of approximately 5 to 10μm, which justifies the use of the best microelectronics technology available. Figure 17 lists the characteristics of some precision thin-film resistors.

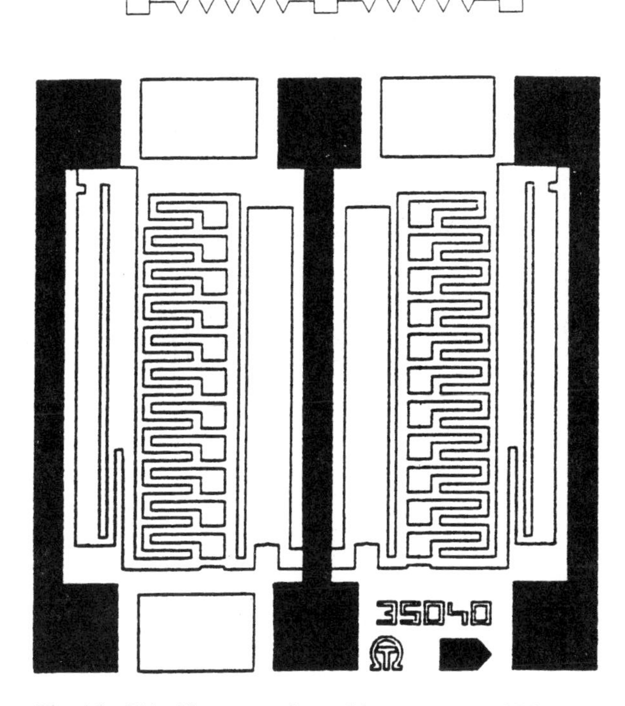

Fig. 16 Thin-film two resistor chips, center tap, 075mm square.

PRECISION THIN FILM TECHNOLOGY

Silicon, Standard Resistor Arrays

RMK 408N, 508N, 48N, 816N and 914N Series

Manufactured in ULTRAFILM® technology, these resistor network chips have a high level of integration, wide ohmic value range, very low temperature coefficient which are unequalled on the market today. Laser trimming can provide excellent precision.

Features

- High precision tolerances
- Very low temperature coefficient
- Excellent stability

Typical Performance

■	ABS	TRACKING
TCR	5	1
	ABS	RATIO
TOL	0.1	0.01

Fig. 17 Specifications for thin-film arrays (manufactured by Sfernice, a division of Vishay).

6. HIGH PRECISION LOW-OHMIC VALUE RESISTORS

Low-ohmic value resistors include those with ohmic values between 3 and 0.001 ohms. These resistors must also display the same characteristics as other high-precision resistors.

Figure 18 is a schematic of a low-ohmic value resistor. Since the leads have significant resistance, the incoming current is independent of the actual signal that causes the potential drop $V_1 - V_2$ *across the terminals of the active part of the resistor.*

Four wires are used: two for the current and two for the signal. In practice the resistances, r_c, include not only the resistance of the leads but the resistance of the solder between the leads and the active part of the circuit. The electrical properties of the leads and solder (*TCR,* stability) are generally compatible with the desired performance of the component.

It is up to the manufacturer to reduce the resistivity of the solder and any related dispersion so that the tolerance, *TCR,* and stability of the component are well within the limits of the customer's specifications or the performance claimed by the manufacturer.

The primary resistance element is created by arranging several resistors in parallel. Of the various technologies described in this chapter, only foil technology can be used to manufacture such components. Wirewound resistors cannot be used for this purpose, even though they can be made from wires arranged in parallel. The problem is that the tolerances associ-

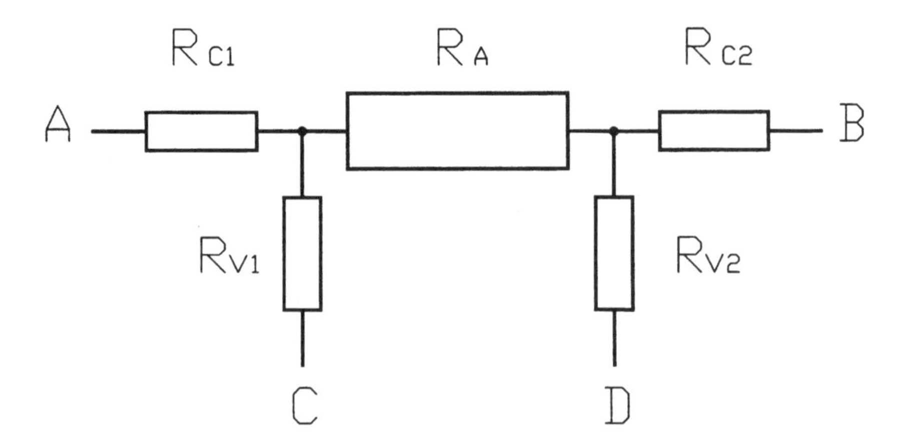

Fig. 18 Four wire (Kelvin) connection of a low-ohmic value resistor.

ated with wire length adjustment or the process of chemical or mechanical erosion are very difficult to maintain. Additional problems arise in trying to control resistance dispersion in the solder.

Thick-film (cermets) or thin-film materials are not necessarily well suited to the production of low-ohmic value high-precision resistors. Low-ohmic value pastes, with values below $10\Omega/\blacksquare$, are unstable. The same is true of Ni-Cr or Ta thin-film materials.

Foil technology, however, is well suited to the production of such devices. It is relatively easy to manufacture foils only a few micrometers in thickness, with a resistivity between 0.5 and 0.15 $\Omega/\blacksquare$. These foils are etched by chemical or electrochemical means to obtain parallel resistor circuits containing adjustment points as described earlier. The contacts between the input current and the actual signal are built up from electrolytic copper deposited on the foil.

Specific techniques can be used to minimize the resistance of the interfaces as well as any associated dispersion.

Figure 19 represents the schematic for a low-ohmic value foil resistor. The input current and measurement pads, along with the adjustment points, are clearly visible in the illustration.

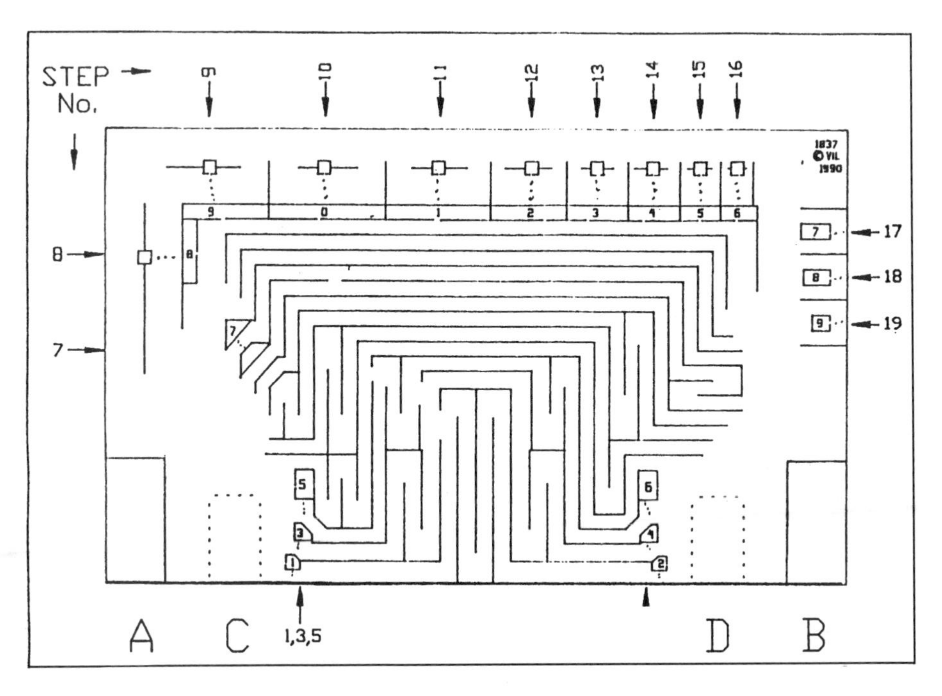

Fig. 19 Pattern of low-ohmic value foil resistor.

By using three foil thicknesses (2.5, 5, and 10μm), the resistor can be adjusted to any value between 0.1 and 1Ω, with an accuracy of a few ppm and *TCR* of ±5ppm/°C. Very thick foils can also be used. In this case there is no need for a ceramic substrate as the foil is rigid enough. The adjustment of value is done with a laser, and the terminals consist of welded copper wires or strips (Chap. 9, Fig. 8b). In this case values down to 1 milliohm have been achieved.

7. THERMOELECTRIC EFFECT

The thermoelectric effect, which is negligible in ordinary resistors, may become a significant source of drift or instability in high-precision resistors.

Known as the Seeback effect, it occurs when the following two conditions are present at the same time:

1. An electrical circuit is made from two different conducting materials (metals M1 and M2), which are soldered at their ends, A and B (Fig. 20).
2. A temperature difference $T_2 - T_1$ exists between A and B.

Solution of the Boltzmann transport equation for solids shows that the electrical and thermal currents depend on the electric field and temperature gradients:

$$\vec{J} = L_{EE}\vec{E} + L_{ET}\,\overrightarrow{\text{grad}}(T^\circ) \qquad (1.14)$$

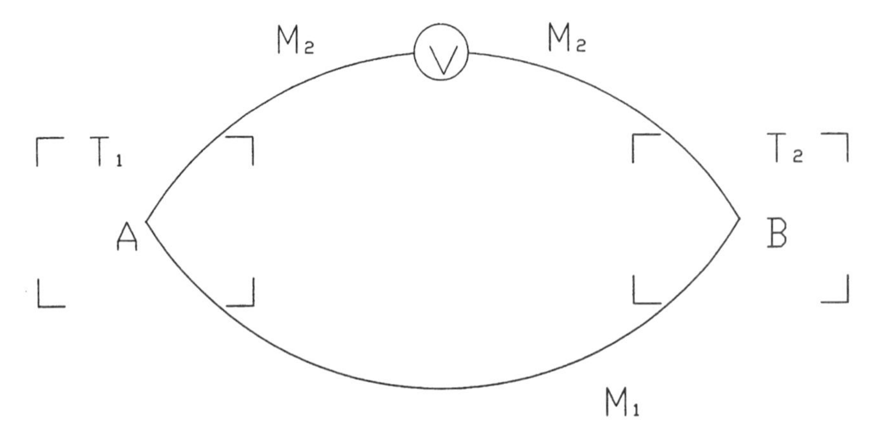

Fig. 20 Seeback effect.

$$\vec{Q} = L_{TE}\vec{E} + L_{TT}\,\overrightarrow{\text{grad}}(T°) \tag{1.15}$$

Here, $\vec{J}$ and $\vec{E}$ are the current density and electrical field in the conductor, and $\vec{Q}$ and $\overrightarrow{\text{grad}}(T°)$ are the thermal current density and temperature gradient between the ends of the conductor. The coefficients L depend on the nature of the conductor.

In practice we cannot observe these coefficients directly. If the specimen is at constant temperature, however, then $\vec{J} = \sigma\vec{E}$ and then $\sigma = L_{EE}$. If the specimen is electrically insulated in such a way that no current circulates and temperature gradient is maintained between the ends, then $\vec{Q} = K \cdot \overrightarrow{\text{grad}}(T°)$ since $\vec{J} = 0$, $\vec{E} = L_{ET}/L_{EE} \cdot \overrightarrow{\text{grad}}(T°)$ and:

$$K = -[L_{TT} - L_{ET}L_{TE}/L_{EE}] \tag{1.16}$$

The ratio $S = -L_{ET}/L_{EE}$ is called the thermoelectric power of the conducting material.

In Fig. 21 the measurement voltmeter is at an intermediate temperature $T_1 < T_0 < T_2$. The Seeback voltage between two conductors $M1$ and $M2$ is then given by the contour integral.[8] $TH.EMF = -\int \vec{E}\,\vec{dr}$, which can be written:

$$TH.EMF = -\int S \cdot \frac{\partial T}{\partial r}\,dr = \int_{T_0}^{T_1} S_B dT + \int_{T_1}^{T_2} S_A dT + \int_{T_2}^{T_0} S_B dT$$

$$= \int_{T_1}^{T_2} (S_1 - S_2)dT \tag{1.17}$$

Here, r is the contour of the circuits A and B.

Since the output connectors are generally different from the material used for the resistor, an EMF may occur either from external sources of heat or from heat gradients generated when electrical power is applied to the resistor, or both simultaneously. The equation above shows that two conditions are required to minimize $TH.EMF$:

A. Select connector materials in which $S_1 - S_2 \cong 0$.
B. Design the circuit so that $T_2 - T_1 \cong 0$.

In discrete wirewound or foil resistors, whose volume is usually around 20–100 mm^3, condition B is difficult to realize in all cases and manufacturers generally limit themselves to selecting materials that fulfill condition A.

[8]Ziman, J. M. *Electrons and Phonons,* p. 271. Oxford University Press, 1963.

The dimensions and assembly methods (surface-mount technology) used to produce thin-film resistors and resistor networks, however, lend themselves well to satisfying condition B. The choice of interconnection materials is not a critical factor and other criteria (mechanical and electrical) are used in their selection. Manufacturers always indicate in their catalogs the *TH.EMF* values of their products. Some typical orders of magnitude are: 3 to 10μV/°C for wirewound resistors, 0.002 to 1μV/°C for foil resistors, and 0.002 to 0.005μV/°C for thin-film and network resistors.

Since the *TH.EMF* is particularly to be avoided in very low-value resistors, and foil resistors are best for low values, they must be very carefully designed to avoid this effect.

8. COMPARISON OF THE DIFFERENT TECHNOLOGIES

Table 3 compares the different technologies used to manufacture precision resistors.

9. CONCLUSION

The electronics industry has developed extremely sophisticated methods for fabricating high-precision resistors designed for use in measurement devices, computers and communications systems. These components display remarkable electrical characteristics and, with the exception of wirewounds, which become bulky, are well suited to surface-mount technology. At present, there is no comparable technology in resistor manufacturing that appears likely to provide similar performance.

TABLE 3 Comparison of Different Technologies used for Production of Precision Resistors

	Wirewound	Foil	Cermet	Thin Film
Resistance in Ω	0.1Ω–55MΩ	0.1Ω–150KΩ	0.1Ω–15MΩ	10Ω–3MΩ
TCR ppm/°C	1–25	0.5–10	15–300	5–25
Δ*TCR* (network) ppm/°c	0.5–5	0.1–3.0	15–20	2.0
Integration (network)	yes, but difficult	excellent	yes	excellent
Stability at 125°C and 10,000 hr. ppm	150 typical	150 typical	.2%–.5% (2000 hrs @ 70°C)	500–1000 (2000 hrs @ 70°C)
Relative stability ppm (network)	100ppm/yr	200 75 typical	.05%–.4%	<300
1/f noise	low	low	high	low
Frequency response	mediocre	excellent	good, but noise present	excellent
Reliability	excellent	excellent	excellent	excellent
Cost	very high	high	low	medium
Surface-mount	yes	yes	yes	yes

The specifications for high-precision resistors list a number of factors that must be taken into consideration during production:

Resistance value and adjustment accuracy
Temperature coefficient
Dissipated power
Dimensions
Contacts (wire, ball-bonding, ultrasonic bonding, etc.)
Stresses caused by the surrounding circuit
Operating frequency
Moist heat capacity
Integration density for networks

On the basis of these specifications, the manufacturer will suggest a specific technology, wirewound, thick film, foil, or thin film, for the intended application.

11

Potentiometers and Potentiometric Position Sensors

1. INTRODUCTION

Electronic circuits contain amplifying devices used to adjust the signal level depending on the characteristics required by the user. This is particularly true for analog and measurement electronics. A person listening to a stereo system can adjust the sound volume by modifying the amplitude of the source signals fed to the amplifiers. The measurement of a physical phenomenon by means of an electronic device takes place in such a way that the amplitude of the input signal representing the phenomenon is modulated as a function of the characteristics of the measurement circuits. This modulation can take place through the use of a potentiometer. We can represent a component of this type schematically by a voltage bridge in which V is the input signal and v the output signal:

$$\frac{v}{V} = \frac{R_1}{R_1 + R_2}$$

Here, $R_1 + R_2 = R_T$ is constant while R_1 and R_2 vary continuously.

Under other circumstances the change in R_1 can be linked to the movement of an object, such as a piston; the position of an air intake valve on a carburetor; the expansion of a mechanical part with temperature, or any other limiting factor. In this case we speak of a potentiometric position sensor. Below, we discuss the design of such components, their performance, and their operating conditions.

2. POTENTIOMETERS AND TRIMMERS

There are two main types of variable resistors: potentiometers and trimmers. Whereas potentiometers are generally devices that respond to variations in an input signal in a circuit or electronic amplifier (an audio amplifier, for example), trimmers are used to tune electronic circuits during their development by adjusting voltage level needed to power a circuit or the output voltage of an operational amplifier. Once the circuit has been tuned, the position of the trimmer generally remains unchanged. This leads to notable differences in their characteristics. The potentiometer, which is often adjusted as a function of signal amplitude, must have a lifespan that can sustain thousands (and, with plastic tracks, millions) of operations without noticeable loss in performance. Trimmers, on the other hand, can have lifespans of hundreds of operations. Since trimmers are often used as variable resistors (rheostats), their TCR should be lower than that for potentiometers. As a result we find commercial trimmers, that for special applications, are manufactured from Ni-Cr foils possessing a TCR on the order of ±5ppm/°C[1] (see Chap. 10 for a full description of the use of this material). In general commercial trimmers have a TCR of 50ppm/°C or more. In the remainder of this chapter, we will refer to both types of devices as potentiometers.

Potentiometer construction is very straightforward. It consists of a resistor R (called the track) fed by a voltage V, and a voltage tap that slides along this resistor, receiving a voltage v proportional to its position along the resistor (Fig. 1). If the cross-section and resistivity of the resistance R are constant, the ratio v/V is thus equal to:

$$v = V \cdot \frac{l}{L} \tag{1.1}$$

Potentiometers can be rotary or linear. Figure 2 shows the structure of a rotary and a linear potentiometer.

[1]See the Vishay catalog, for example.

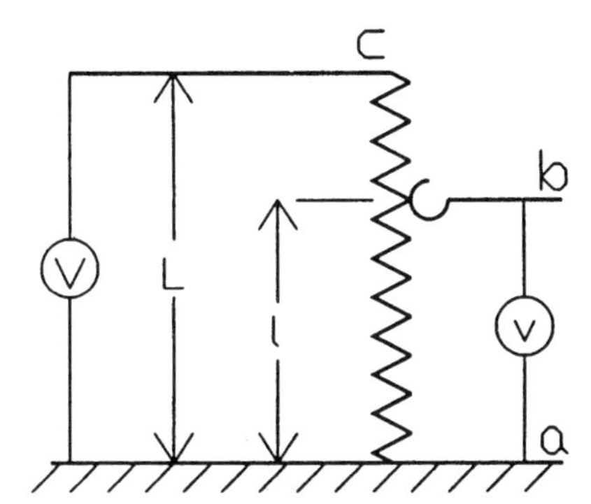

Fig. 1 Schematic of a potentiometer.

2.1 Track Materials and Characteristics

Potentiometer tracks in general consist of two types of material:

- Cermet films
- Plastic films

Some are also made of wire winding on a mandrel or of foil bonded to a substrate. Cermet films do not differ in their physical properties from similar materials used in fixed resistors. They use the same pastes with the same characteristics in terms of TCR, stability, and longevity. Their surface finish is such that they are not (overly) abrasive for the wiper contacts. Ohmic contacts are also made of conductive silver or silver-palladium conductive cermets. These potentiometers are used in industrial, professional, and military applications, in which severe operating conditions and the need for reliability require the use of specialized materials.

For consumer applications (household appliances, radios, stereo systems, etc.), the tracks consist of a mixture of graphite and prepolymer that is premolded and shaped on a phenolic substrate. The ohmic contacts are made from silver paste. Figure 3 represents some typical potentiometers.

2.2 Wipers: Characteristics and Materials

The wiper is designed to provide an electrical connection with the resistance track in such a way that the electrical response of the system is a function of the wiper position. The contact moves, therefore, and exerts a

Bulletin 1R-600

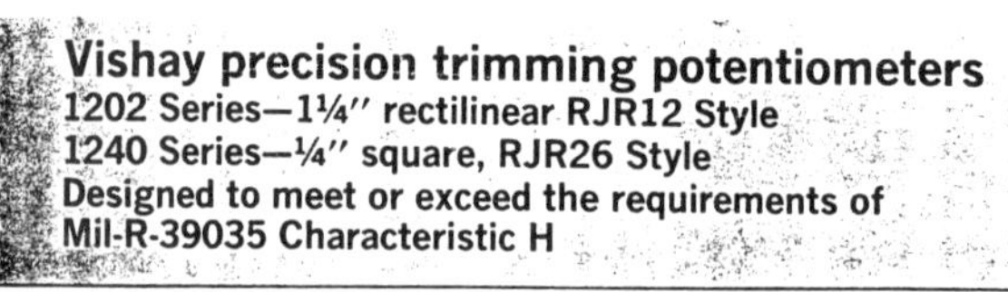

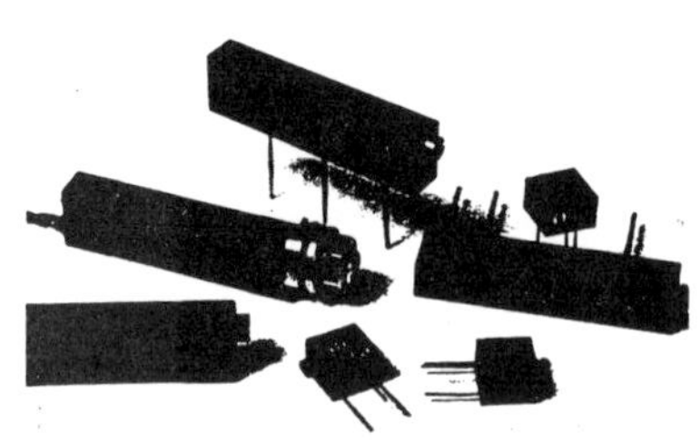

- **TCR** 10ppm/°C—full Mil range
- **Equivalent Noise Resistance (zero-based CRV)** 10Ωmax
- **Settability** 0.05%
- **Hop-off** Unmeasurable
- **Stability (load-life)** 1.0% ΔR max @ full rated power for 2,000 hours (1202), for 10,000 hours (1240)
- **Resistance range** 2Ω to 20kΩ

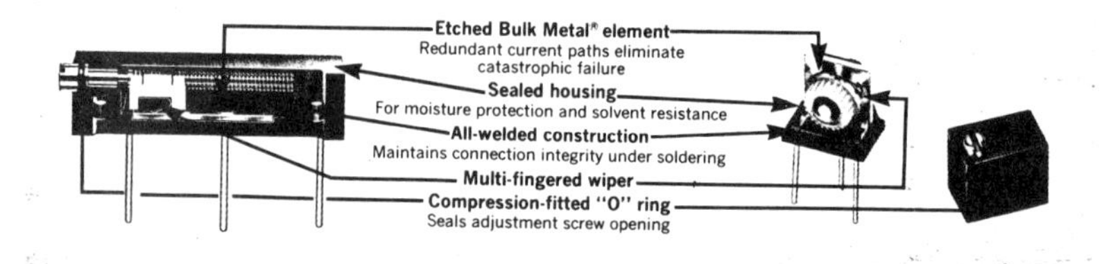

3

TABLE I—Standard ordering information

1202	Y	1k	5%
MODEL	**TERMINATION STYLE**[1]	**STANDARD RESISTANCE VALUE (ohms)**	**TOLERANCE**
1202 RJR12 style	Y[2] —Staggered PC pins P —In-line PC pins L —Flexible wire leads LB —Flexible wire leads with bushing	2, 5, 10, 20, 50, 100, 200, 250, 500, 1k, 2k, 2.5k, 5k, 10k, 20k	5%, 10%[3]
1240 RJR26 style	P —Horizontal-mount, side-adjust W —Edge-mount, top adjust X —Edge-mount side adjust	5, 10 20, 50, 100, 200, 500, 1k, 2k, 5k	10% 5%

1. See appropriate termination style drawing for exact dimensions, pinouts, and mechanical details.
2. Preferred termination style for current 1¼" rectilinear designs. (Staggered PC pins present a sturdier arrangement for shock, vibration, and impact situations.)
3. Standard tolerances for the 2Ω, 5Ω, and 10Ω units are 10% and 20%. A 5% tolerance for these values is available on special order.

Fig. 2 Construction of linear and rotary trimmers.

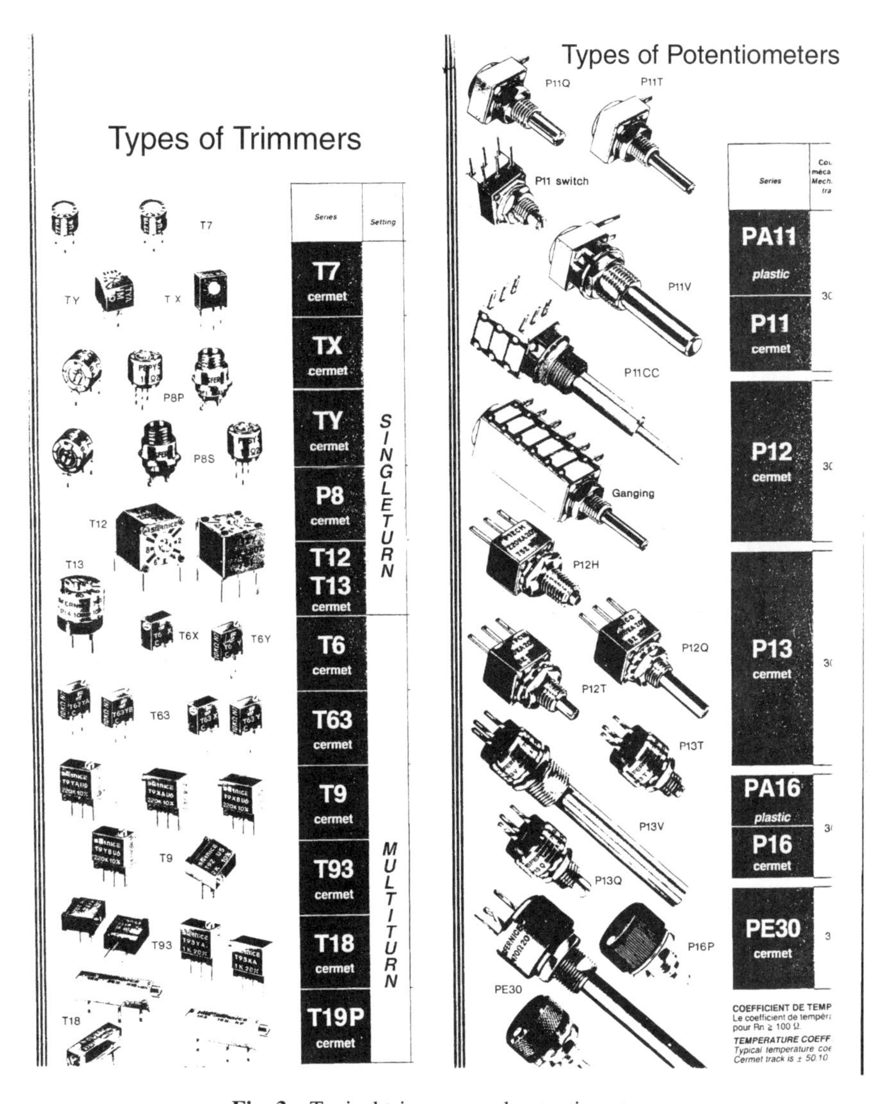

Fig. 3 Typical trimmers and potentiometers.

certain pressure (as small as possible so that the track and wiper do not wear out too quickly). In doing so it creates a contact resistance, which is described briefly below.

2.2.1 Contact resistance[2]

If we have a conductor of resistance R, we can make a cut parallel to its cross-section and press the two surfaces created against one another. The resistance is now $R' > R$. The additional resistance is the contact resistance $R_C = R' - R$. A fuller analysis shows that this resistance is formed of two independent components $R_C = R_F + R_S$.

R_F is the resistance created by the film of foreign matter found in the ambient and absorbed at the surface of the contacts, or from a chemical reaction that occurs with the material forming the contacts (oxidation, for example). This resistance is located between the contact elements.

R_S is the so-called "constriction" resistance (see Chap. 6, section 4.5), which arises from the fact that the contact is effective on only a very small fraction of the apparent surface, and the lines of current must converge toward the points of flow. The resistance associated with this convergence is located within the elements that are in contact.

By considering a single point of contact, the resistance R_c can be broken down as shown in Fig. 4.

Assume we have a disk of radius a whose thickness is negligible and whose resistance is, therefore, zero. It represents the point of contact between materials 1 and 2 of identical resistivity ρ (Fig. 5). The equipotential points are flattened ellipses that become circles when the distance is

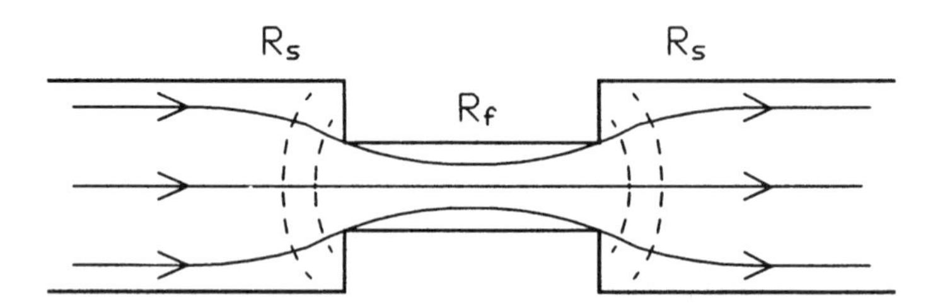

Fig. 4 Resistance in a single point of a contact.

[2]Cherbuy, J., Personal correspondence. Vishay–Sfernice Research Department.

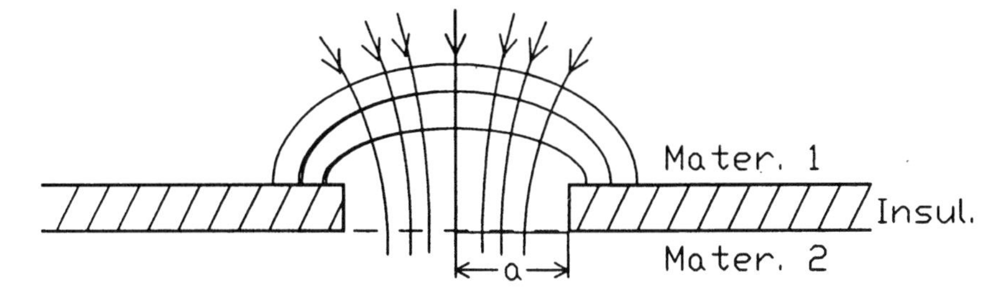

Fig. 5 Equipotential lines in a point of contact.

great. The solution to Laplace's[3] equation (this is similar to the example given in Chap. 4) gives:

$$R_c = \frac{\rho}{2 \cdot a} \tag{1.2}$$

In practice there is not just a single point of contact, but many, distributed randomly over the exposed contact surface. If these contact points are not too close together, so there is no interference from current lines, the total resistance per unit surface will be equal to:

$$\frac{1}{R_{ct}} = \frac{2}{\rho} \sum_{1}^{n} a_n \tag{1.3}$$

in which n is the average number of contacts per unit surface.

This resistance, R_{ct}, will vary with the applied pressure P. The elastic deformation caused by the pressure of the contacts on one another will cause a to vary. Without going into the details, for a given group of contact materials:[4]

$$R_S \propto \rho \cdot P^{-c} \qquad 0.3 < c < 0.5 \tag{1.4}$$

The passage of a current will heat the contact. This temperature T_{max} can be estimated by measuring or calculating the potential difference U along the constriction. Taking the tube of current shown in Fig. 5, we see

[3]Holmes, R., *Electric Contacts, Theory and Applications.* Springer-Verlag, 1950.

[4]Holmes, R., *op. cit.*

that the spatial distribution of the electric and heat potentials obeys the same Laplacian equation. The same amount of heat is produced everywhere on both sides of the contact, which thus experiences no heat flow. The energy $I \cdot v$ (I being the current crossing the tube) produced between the equipotentials 0 and $-v$ flows only through $-v$. The flow (electrical and thermal) produces a potential difference when it crosses an electrical or thermal resistance held between two neighboring equipotential values. Consequently,

$$-dT = I \cdot v \cdot dR_{\text{th}} \text{ (Fourier's law)}$$

$$dv = I \cdot dR \text{ (Ohm's law)}$$

$$-\frac{dT}{dv} = v \cdot \frac{dR_{\text{th}}}{dR}$$

Since the temperature and potential fields are solutions of the same Laplacian equation, the geometry of the resistors is the same. As a result,

$$\frac{dR_{\text{th}}}{dR} = \frac{\sigma}{k} \quad \frac{\text{electric conductivity}}{\text{thermal conductivity}}$$

and

$$-\frac{k}{\sigma} \cdot dT = v \cdot dv$$

There is an equation known as the Wiedemann-Franz law that states for reasonably high temperatures (300°K to 700°K) the ratio of thermal and electric conductivity is directly proportional to the temperature, so that $k/\sigma = L \cdot T$, where L is the Lorenz number ($2.45 \cdot 10^{-8}$ W · ohm/deg^2).

If we assume that this equation can be applied to the materials used in sliding contacts, then:

$$-L \cdot T \cdot dT = v \cdot dv$$

By integrating over the current half-tube, we get:

$$-L \cdot \int_{T_{\max}}^{T^\circ} T \cdot dT = \int_0^{v_0} v \cdot dv \quad \text{and, therefore,} \quad -L \cdot \frac{T_0^2 - T_{\max}^2}{2} = \frac{v_0^2}{2}$$

v_0 is half the value of the potential difference we can measure between two points along the constriction. Consequently, $v_0 = U/2$ and

$$T_{\max}^2 - T_0^2 = \frac{U^2}{4L} \tag{1.5}$$

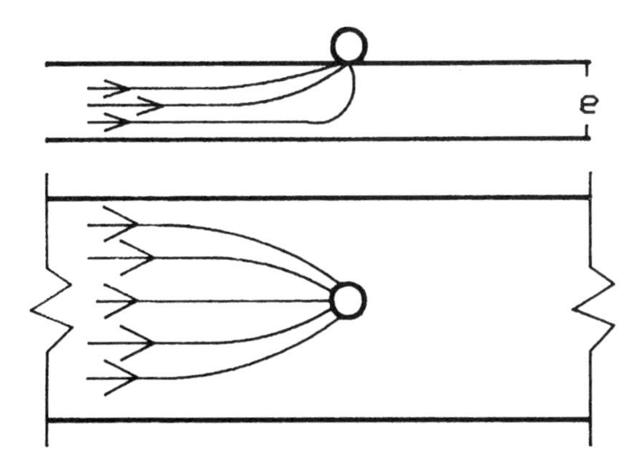

Fig. 6 Constriction resistance.

The value of U can be estimated by calculating the equipotentials at the constriction.

This effect, which is ordinarily negligible for potentiometers, can be a source of problems and instability in the trimmers used in rheostats. For example, if $U \cong 0.1\text{V}$, T_{max} can reach 300°C locally.

Compared to the effects above, the contact of a wiper against a track displays some specific characteristics. The resistivity of the track is always one or two orders of magnitude greater than that of the wiper. As a result, the constriction resistance is primarily found in the track. We can represent this schematically by lines of current along the track, viewed from above and from the side, as shown in Fig. 6.

The above drawing, in which the current passes through the wiper, can be represented by the circuit shown in Fig. 7.

The very high-impedance voltmeter measures the potential drop $I \cdot R_c$, by which we can measure R_c, regardless of the wiper position.

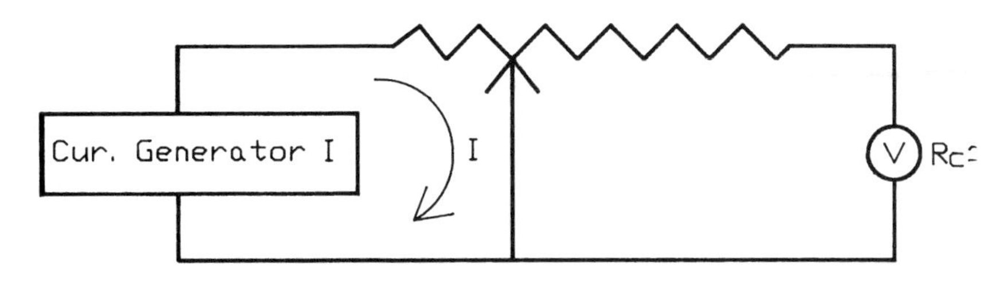

Fig. 7 Measurement of contact resistance.

2.2.2 Potentiometer noise

When we activate the wiper in the preceding device, we obtain a measurement of R_c that varies as a Gaussian noise (Fig. 8). The amplitude of this noise can reach values on the order of 10^{-3}—10^{-2} ohm.

Both the distribution of the effective points of contact between the wiper and the track, and the size of radius a, the latter related to the microscopic condition of the surface, vary from one position to the next.

Track resistivity also changes randomly due to the heterogeneity of the materials used (cermet and graphite). These two factors cause fluctuations in the constriction resistance. Finally, foreign materials (dust grains, waste produced by wiper and track wear) and surface irregularities constitute random R_F resistances between the wiper and the track. This last phenomenon is the most often cited, mistakenly, as the cause of the noise. We[5] have been able to show, in fact, that the contact noise of tracks with low-ohmic values—that are thus highly charged with conductive grains and consequently very rough—had a lower RMS than that caused by tracks with elevated ohmic values and highly variable local resistivity but smoother surfaces.

2.2.3 Wear

As indicated above, a potentiometer must accomplish more than a few operations: Lifespan specifications demand several thousand operations for small potentiometers and several million for displacement transducers or precision potentiometers.

Under these conditions, we need to use contact blades that are sufficiently stiff and pressures that are sufficiently weak to avoid premature wear of the track and/or the wipers. The materials used for this purpose are

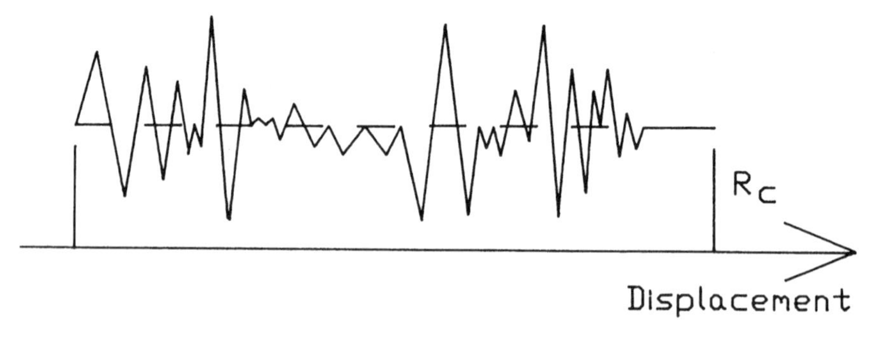

Fig. 8 Noise resistance.

[5]Vishay–Sfernice, Direction Recherches. Private communications.

Table 1 Types of Wipers Used in Practice

Potentiometers	Trimmers	Precision Potentiometer
Carbon grain	Paliney 6[a]	Paliney 6
	Copper-Beryllium Alloy	Paliney 7
	Gold-plated Palladium	(for high performance)

[a]Paliney alloys get their name from the company that manufactures them.

Paliney 6: % of atomic mass, Pd (44%), Ag (38%), Cu (16%), Pt (1%), Ni (1%); average lifespan 20–30 $\times 10^6$ cycles

Paliney 7: Pd (35%), Ag (30%), Cu (14%), Au (10%), Pt (10%), Zn (1%); average lifespan $>100 \times 10^6$ cycles

copper-aluminum alloys (in variable proportions) or copper-beryllium alloys for traditional potentiometers, and ternary alloys of the copper-palladium-silver or gold-palladium-platinum type for precision potentiometers and displacement transducers (see Table 1).

As can be seen from the above, the conditions required for a good wiper/track contact are the opposite of those needed for low wear.

This problem is solved by using multiple blades (Fig. 9). For a total given pressure, each contact finger, weakly charged individually, has a

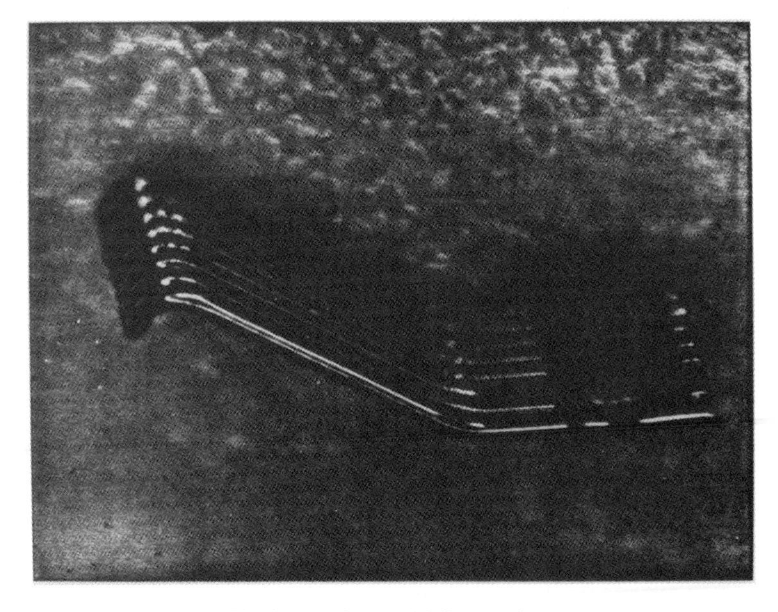

Fig. 9 Typical multi-finger wiper.

FEATURES

- Designed for efficient, accurate miniaturization
- Can be wave or dip soldered without rotor problems
- Coded marking for easy identification of resistance value
- Models for standard, automatic or reverse adjustment
- 12mm tape and reel packaging

ELECTRICAL SPECIFICATIONS

Resistance Values: 100 ohm to 1 Megohm.

Resistance Tolerance: ±30%.

Rated Power: 0.2 watt at 70°C.

Rated Voltage: 100 VDC.

Rotation Life: 20 rotations.

Operating Temperature Range: −40°C to +100°C.

TCR: ±250PPM/°C.

CRV: 5% maximum.

Fig. 10 Characteristics of potentiometers.

high contact resistance but produces very little wear of the track. Total contact resistance comes from setting all the fingers in parallel, reducing the noise variance.

Besides the traditional components, miniaturized surface-mounted potentiometers and trimmers are now available. Figure 10 summarizes the technical characteristics of all the component families of this type on the market.

3. DISPLACEMENT TRANSDUCERS AND PRECISION POTENTIOMETERS

Precision potentiometers and displacement transducers are used quite differently from potentiometers and trimmers. These components are enslaved to the displacement or to the position of a mechanical system (axle, piston, vane, level, etc.) in such a way that the v/V ratio provides the position of the mechanical element.

Furthermore, since these components are elements of a control sequence, the displacement of the wiper over the track by a distance x (in the case of a linear potentiometer) or θ (in a rotary potentiometer) follows a mathematical law defined by the user in such a way that:

$$\frac{v}{V} = f(x \text{ or } \theta) \tag{1.6}$$

Consequently, the characteristics and the performance of precision potentiometers are very different from those of ordinary potentiometers.

Consider a potentiometer where the response v/V follows a mathematical law $f(x)$ or $f(\theta)$. *The characteristic function* is given by Eq. 1.6.

Conformity, $\pm C$, is the error between the effective characteristics function given by the potentiometer and the theoretical characteristic function:

$$\frac{v}{V} = f(x \text{ or } \theta) \pm C$$

Absolute conformity (Fig. 11) is defined by the maximum error between the practical characteristic function and the theoretical characteristic function. It is expressed as a percentage of the applied potential V and is measured over all the useful intervals of the curve $f(x \text{ or } \theta)$:

$$\frac{v}{V} = f(x/x_T \text{ or } \theta/\theta_T) \pm C; \quad 0 \leq x \leq x_T \text{ or } 0 \leq \theta \leq \theta_T$$

If $f(x \text{ or } \theta)$ is a linear function, this will be referred to as *absolute linearity.*

Resolution is the measurement of the sensitivity with which a given value of the ratio v/V can be established.

Noise is given by the ratio of the variations in v due to contact resistance and the reference voltage V. This parameter integrates all the effects of contact resistance variations, resolution and other nonlinearities for a given wiper position. All these definitions and the testing methods used to measure these effects are well standardized by the manufacturers.[6]

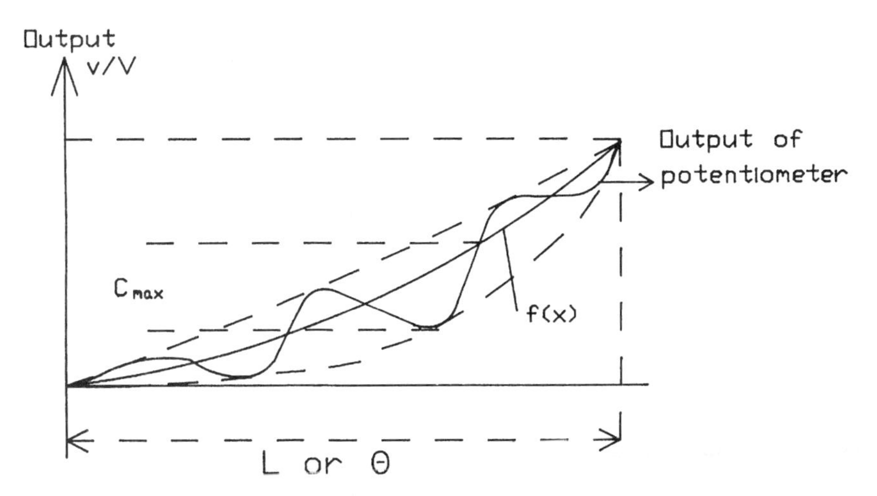

Fig. 11 Absolute conformity.

[6]For more details refer to *Industry Standards Wirewound and Nonwirewound Precision Potentiometers*, Variable Resistive Components Institute, P.O. Box 1070, Vista, CA 92085-1070.

Besides these characteristics, precision potentiometers and motion transducers have a lifespan of many million operations (about 10^7–10^8 operations) coupled with a rotational speed or a linear wiper speed over the tracks which is one or two orders of magnitude superior to those accepted for conventional potentiometers.

4. PRACTICAL RESULTS

These working conditions impose a certain number of constraints on the structure of precision potentiometers. The fast rotational or displacement speeds call for the use of precision ball bearings.

Commonly accepted standards demand absolute conformity to within 0.01%. This translates in practice into the need to adjust the component's response as much as possible to the computer-defined $f(x \text{ or } \theta)$ standards.

A lifespan of several million operations requires very advanced technologies in the production of tracks and wipers. This is all the more so because the level of noise must not exceed 0.01% of the full-scale response.

Figure 12 shows assemblies made up of a rotary precision potentiometer and a linear potentiometer.

4.1 Nature and Structure of the Tracks

Wirewound precision potentiometer tracks are still in use today. Nevertheless, the current tendency is to use so-called "plastic" tracks. For good reason: Their manufacturing cost is lower, the wear is less, and this technology is better suited to "adjustments" that fit the component's outgoing signal within the required absolute conformity. Plastic tracks are made up of a polymer and carbon black mixture deposited on a plastic surface.

The polymer is chosen for its mechanical properties and to support the numerous wiper movements without notable wear. Two technologies are used:

- Using appropriate masks, the polymer-carbon mixture is deposited on a diallyl phtalate surface and the resulting component is "molded together" into the final shape and dimensions chosen for the potentiometer. This technology is generally used for standard potentiometers of small dimensions and relatively low cost.

- The polymer-carbon mixture is deposited on a plastic teflon sheet or any other type of sheet likely to provide the stability conditions required

Fig. 12 Linear and rotary motion transducers.

by the mechanical, electrical, and thermal constraints. The ink is then polymerized under conditions imposed by the nature of the materials used. Large sheets are thus obtained where tracks will be cut mechanically and pasted in the required places within the potentiometer structure. The silver or gold polymer paste electrical contacts are then attached at the points determined by the potentiometer's structure. This technology will be used primarily to manufacture potentiometers with very specific $f(x \text{ or } \theta)$ responses (for example, $f(x \text{ or } \theta)$ will be power series functions in x, exponential functions, etc.), and for very large potentiometers (some displacement transducers for controlling molding injection rams are one to two meters long).

Manufacturers will use their know-how to obtain tracks that display homogeneous resistivity and carbon particle dispersion.

Fig. 12 (*continued*)

4.2 Wipers

The wipers will be made of precious metal alloys of the copper-palladium-silver or gold-palladium-platinum types (see their characteristics in Table 1), which provide both hardness and perfectly smooth surfaces in order to obtain minimal track wear. Precious metals are chosen because the surfaces will then be free of oxidation. Figure 13 shows the typical wiper structure.

4.3 Adjustment

Potentiometers whose conformity is of the order of $\pm 1\%$ to $\pm 0.5\%$ generally do not need adjustment (trimming). All other potentiometers are adjusted.

The operation is performed with a device that includes a computer that acquires voltage-response and wiper-position data. For one given position of the wiper, in x or in θ, the computer calculates the theoretical value of $f(x)$ or $f(\theta)$ and compares it, by means of a voltage meter and an A/D converter, to the actual value provided by the potentiometer. A plotter provides the value of the error e, while the system controls a laser source to abrade the appropriate regions of the track edges to make the response comply with the specifications (Fig. 14).

4.4 Nonlinear Laws

The procedure used to manufacture the tracks whose response is a nonlinear function $f(x)$ is based on the solution of Laplace's equation for complex variables.

In fact, assume a complex function $W(z)$. Since $z = x + iy$, the expression can be written as $W(z) = U(x, y) + iV(x, y)$.

Fig. 13 Typical wiper in a motion transducer.

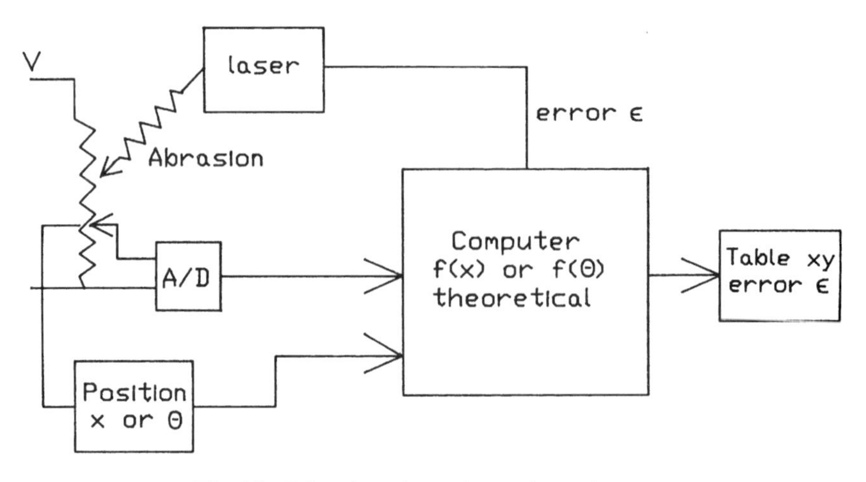

Fig. 14 Trimming of a track to a close tolerance.

If the derivatives of W in relation to z are unique in the x,y plane, functions U and V satisfy the Cauchy-Riemann conditions: $\partial U/\partial x = \partial V/\partial y$ and $\partial U/\partial y = -\partial V/\partial x$. Taking derivatives in relation to x and y and adding the results

$$\frac{\partial^2 U}{\partial x^2} + \frac{\partial^2 U}{\partial y^2} = \frac{\partial^2 V}{\partial x \partial y} - \frac{\partial^2 V}{\partial x \partial y} = 0 \qquad (1.7)$$

Consequently $U(x, y)$, as well as $V(x, y)$, is one solution of the Laplace equation.

The method of manufacturing a track then becomes relatively simple: Assume there is a track of length L, height l, whose wiper slides over the ordinate line p (Fig. 15).

We propose to use this track to build a potentiometer whose response will be $f(x)$. Let us develop $f(x)$ into an ascending power series of x, stopping at a term such that the error committed is less than a value ε for $x = L$.

$$f(x) = A_0 + A_1 x + A_2 x^2 + A_3 x^3 + \ldots \qquad (1.8)$$

Let us consider the analytical function:

$$f(z) = a_0 + a_1(x + iy) + a_2(x + iy)^2 + a_3(x + iy)^3 + \ldots \qquad (1.9)$$

This function can be broken down into:

$$U(x, y) = R[f(z)] = a_0 + a_1 x + a_2(x^2 - y^2) + a_3(x^3 - 3xy^2) + \ldots \qquad (1.10)$$

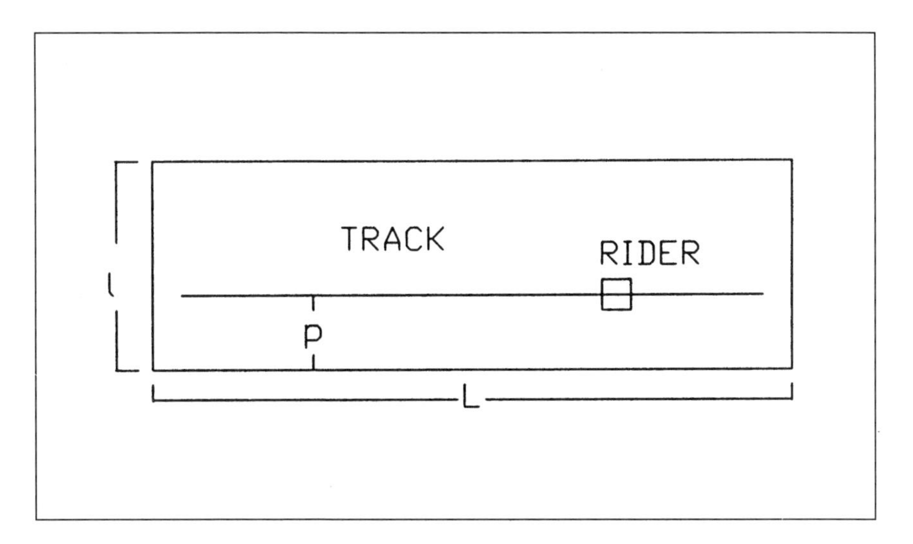

Fig. 15 Track with a rider traveling along an ordinate p.

$$V(x,y) = I[f(z)] = a_1 y + 2a_2 xy + a_3(-y^3 + 3yx^2) + \ldots \tag{1.11}$$

$U(x, y)$ will be the equipotential equation and $V(x,y)$ the equation of the lines of current. The boundary conditions will be set in such a way that $U = 0$ for $x = 0$ and $y = p$, and $U = 1$ for $x = L$ and $y = p$. Consequently, since

$$U(x,p) = a_0 + a_1 x + a_2(x^2 - p^2) + a_3(x^3 - 3xp^2) + \ldots \tag{1.12}$$

by inspection of one by one of the terms between $f(x)$ and $U(x, p)$, it will be possible to determine the parameters a_i as a function of parameters A_i.

Let $a_0 - a_2 p^2 + \ldots = A_0$ (constant term)

$a_1 - 3a_3 p^2 + \ldots = A_1$ (term in x)

etc.

Once these parameters are determined, it will be easy to calculate the equipotentials $U(x, y)$ and the lines of current $V(x, y)$, and to determine the points where they will intersect $y = l$.

For example, we are going to calculate a track whose very simple response is $f(x) = Ax^2$ and whose total resistance is $R_T = 50\text{K}\Omega$. The dimensions of the track are $L = 20$mm, $l = 5$mm and $p = 2$mm. We have:

$$U(x,y) = a_0 + a_1 x + a_2(x^2 - y^2)$$

$$V(x,y) = a_1 y + 2a_2 xy$$

By inspection, for $y = p$

$$a_0 - a_2 p^2 = 0 \text{ (constant term)}$$
$$a_1 = 0 \text{ (term in } x\text{)}$$
$$a_2 = A \text{ (term in } x^2\text{)}$$

the results are, after several calculations and as a function of the boundary conditions, $U(0, p) = 0$ and $U(L, p) = 1$.

$$U(x, y) = \frac{p^2 + x^2 - y^2}{L^2} \quad \text{and} \quad V(x, y) = \frac{2xy}{L^2} \tag{1.13}$$

Now the main equipotentials and the lines of current can be calculated.

For $y = l$ and $U_0 = 0 \rightarrow x = \sqrt{l^2 - p^2} = 4.58\text{mm}$ (point A)

This equipotential intersects the line of current having a value of:

$$V_1 = \frac{2l}{L^2} \cdot \sqrt{l^2 - p^2} = 0.11456$$

Now we are going to calculate a line of current V_2 intersecting $x = L$ such that $y = p$.

$$V_2 = \frac{2p}{L} = 0.2$$

This line of current will intersect the boundary $y = 5\text{mm}$ at point B such that

$$\frac{2p}{L} = \frac{2lx}{L^2} \rightarrow x = \frac{pL}{l} = 8\text{mm}$$

At this point, the value of the equipotential is equal to $U_2 = 0.1075$.

We are now able to establish the track's geometry point by point.

The total value of the resistance R_T is such that $U_1 = R_T V_2$; that is, taking $U_1 = 1V$ and $V_2 = 0{,}2\ A$, $R_T = 5\Omega$. The specifications indicate a total resistance of $R_T = 50\text{K}\Omega$. The track must then be manufactured using an ink of sheet resistance $\rho\blacksquare$ (ohms/square) such that:

$$50\text{K}\Omega = 5\Omega \cdot \rho\blacksquare \quad \text{or} \quad \rho\blacksquare = 10\text{K}\Omega/\blacksquare$$

Finally the equipotential $U_2 = 0$ intersects the line of current V_2 at point C of coordinates ($x = 6.48\text{mm}$, $y = 6.16\text{mm}$).

To reproduce the derived expression exactly, the potentiometer must maintain the shape of the equipotential field and the lines of current, defined by the solution of the Laplace equation and represented by Fig. 16.

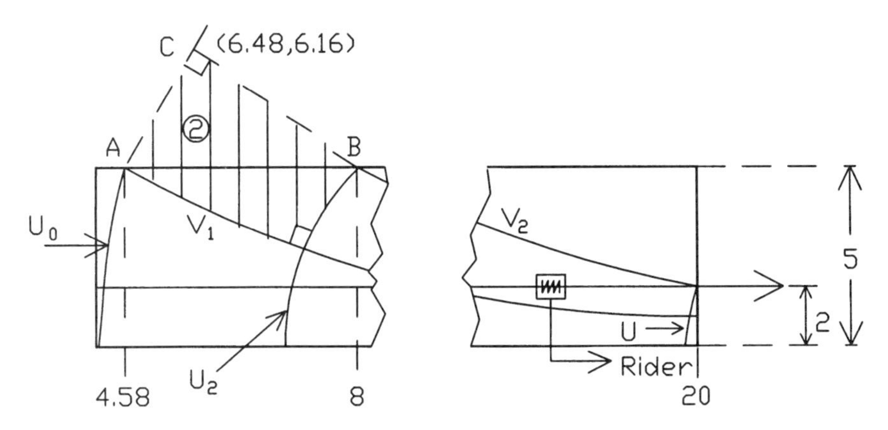

Fig. 16 Computed form of the field.

Unfortunately, the height is limited to the value of 5mm. This makes it necessary to remove part 2 of the track and replace it with a resistance of the same value. With this procedure, the lines of current and the equipotentials of the track will not be modified and will correctly represent the expression in $A.x^2$ on the wiper line.

The value of the resistance corresponding to part 2 is equal to:

$$U_2 = R(V_2 - V_1), \text{ or } R_{ext} = 1.258 \cdot 10k\Omega = 12.58k\Omega$$

The final appearance of the track will then be given by Fig. 17; it indicates the placement of the silvering required to realize the equipotentials U_1 and U_2. The abrasions on the edges represent the lines of current. The resistance R_{ext} can be achieved *in-situ* with the track material or placed on the outside.

We can give another simple example, the fabrication of a sin (x) or cos (x) track. Assume that $f(x) = A \sin(\alpha x)$. Under these conditions $W_{(z)} = A \sin[\alpha(x + iy)] = U + iV$ with:

$$\boxed{U(x, y) = A \sin(\alpha x) \cdot \cosh(\alpha y) \text{ and } V(x, y) = -A \cos(\alpha x) \cdot \sinh(ay)}$$

A central point represents $x = 0$, in order to establish the sine curve over the interval $-L/2 \leq x \leq L/2$

For $x = L/4$, $\alpha x = \pi/2$ and $\rightarrow \alpha = 2\pi/L$

For $x = L/2$, $\alpha x = \pi$, $y = p$ and $\rightarrow V_0 = -A \sinh 2\pi p/L$, then:

$$V_0 = A \sinh(2\pi p/L) - A \cos(2\pi x/L) \cdot \sinh(2\pi y/L)$$

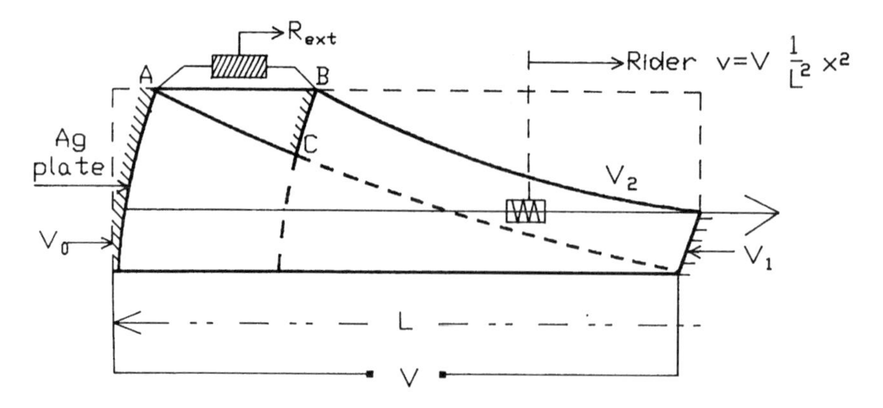

Fig. 17 Actual form of the field.

The last equation determines the entire line of current $V = V_0$. Considering function V at point $y = 1$, where U and V intersect, we can determine a point on the abscissa:

$$x_1 = \frac{L}{2\pi} \cdot \cos^{-1}\left[\frac{\sinh(2\pi p)/L}{\sinh(2\pi l)/L}\right] \tag{1.14}$$

Substituting now x_1 into U_0 for $x = x_1$ and $y = 1$

$$U_0 = A \sinh(2\pi x_1)/L \cdot \cosh(2\pi l)/L, \text{ thus, if } U_0 = 1,$$

$$A = \frac{1}{\sin(2\pi x_1)/L \cdot \cosh(2\pi l)/L} \tag{1.15}$$

As in the preceding case, we now have all parameters to build a sine track (Fig.18).

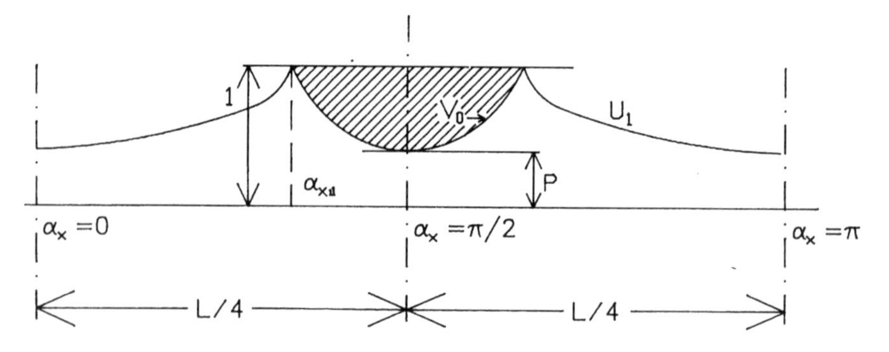

Fig. 18 Sinusoidal track of a rotary potentiometer.

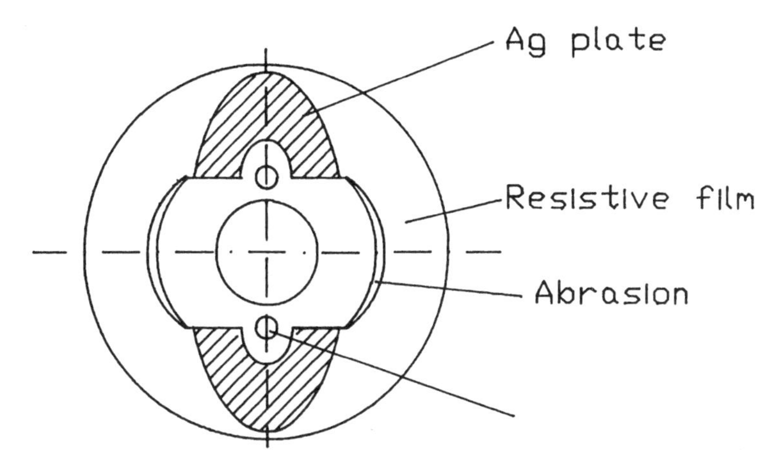

Fig. 19 Geometry of a track per Fig. 18.

The tracks of the potentiometer in $f(\theta)$ are calculated in the same manner. In this case, a system of cylindrical coordinates will be used to represent $U(r, \theta)$ and $V(r, \theta)$. Figure 18 represents the track of a sinusoidal rotary potentiometer.

While the rotary laws will be very complicated, we will have the advantage of making the calculations using conformal transformations methods, after calculating $U(x, y)$ and $V(x, y)$ in the plane, to transform the horizontal and vertical lines into concentric circles and radii. Figure 19 shows the geometry of a sine track for a rotational potentiometer.

5. CONCLUSION

The arrival of digital techniques has curtailed the use of potentiometers, and they are being progressively replaced by automatic digital systems that adapt the signal's entry or exit levels. On the other hand, precision potentiometers and potentiometric transducers are used more and more in automatic control systems, challenging the position of mechanical assemblies. For example, airplane controls currently use such components.

However, for a certain number of applications, they have the disadvantage of producing analog signals that can only be treated by the digital control systems by adding an A/D converter. However, they have the great advantage of not being limited in size, which is not true of their main competitors, the displacement and rotation optical sensors.

12

Magnetoresistance and Its Applications

Anisotropic Magnetoresistance in Ferromagnetic Alloys

1. INTRODUCTION

The phenomenon of anisotropic magnetoresistance in materials and ferromagnetic metals was discovered by Lord Kelvin in 1857, and is an example of a galvanomagnetic effect in conductors. We saw in Chap. 2 that when a conductor is subjected to a magnetic field, its conductive properties are altered and depend on the relative orientation between the electric field (resulting in Ohm's law) and magnetic field. These effects were defined by J. P. Jan[1] for homogeneous isotropic conductors. The author classified them into three types:

1. The magnetic field is parallel to the electric field, and the galvanomagnetic effect is measured in the direction of the current (longitudinal magnetoresistance $\rho_{\parallel}$).
2. The magnetic field is perpendicular to the electric field or current, and its effect is measured in the longitudinal direction parallel to the current (transverse magnetoresistance $\rho_{\perp}$).
3. The magnetic field is perpendicular to the electric field or current, and its effect is measured in the direction perpendicular to the plane

[1]Jan, J. P., "Galvanomagnetic and Thermomagnetic Effects in Metals." *Solid State Physics.* NY: Academic Press, 5 (1957): 1–96.

determined by the electric and magnetic fields. This is referred to as the Hall effect. In metals and alloys this also results in a galvanic effect but it has no practical application because the value of ρ_H is too small.

Longitudinal and transverse magnetoresistance provides a useful means for designing integrated magnetic field sensors.

The magnetoresistance effect is very weak for the metals and alloys currently used in manufacturing conventional resistors. It becomes more significant in nickel monocrystals, as well as in nickel-base polycrystalline alloys and 3*d* transition metals and alloys (the relative difference between $\rho_{\|}$ and $\rho_{\perp}$ is approximately a few percent). Physical models are used to explain anisotropic magnetoresistance.[2] We saw in Chap. 2 that the specific methods of conduction in the transition metals and alloys could be explained by the existence of free energy levels in the *d* band, allowing transitions between the *s* and *d* bands. The high magnetoresistive effect of these metals and alloys can be explained when the spin-orbital interaction in the 3*d* band of these materials is taken into account.

2. DEFINITIONS

In a ferromagnetic metal through which a uniform current flows, $\rho_{\|}$ and $\rho_{\perp}$ are the measured resistivities when the current is, respectively, parallel and perpendicular to the magnetization, whose orientation is assumed to be governed by the application of a magnetic field.

For the majority of ferromagnetic metals, we can show that $\rho_{\|} > \rho_{\perp}$. The anisotropy of the resistivity is defined by:

$$\Delta\rho = \rho_{\|} - \rho_{\perp} \tag{1.1}$$

and the magnetoresistance ratio is defined by the following equation:

$$\frac{\Delta R}{R} = \frac{\rho_{\|} - \rho_{\perp}}{\rho} \tag{1.2}$$

where ρ is the resistivity of a fully demagnetized sample defined by:

$$\rho = \frac{1}{3}(\rho_{\|} + 2\rho_{\perp}) \tag{1.3}$$

[2]Mott, N. H., *Proc. Roy. Soc.* A 153 (1936): 699.
McGuirre, T. R. and Potter, R. I., *IEEE Trans. Magn.* Mag-11 (1975): 1975.

Under isothermal conditions, if we define the magnetic inductance vector $\vec{B}$, then the electric field, due to the extraordinary magnetoresistance, which depends on the direction of the spontaneous magnetic moment in the sample, is given by the following expression:

$$\vec{E} = \rho_{\perp}\vec{J} + \vec{n}(\vec{J}\cdot\vec{n})(\rho_{\parallel} - \rho_{\perp}) + \rho_H(\vec{n} \wedge \vec{J}) \qquad (1.4)^3$$

in which $\vec{n}$ is a unit vector parallel to $\vec{B}$.

We can express the components of the electric field $\vec{E}$ for a ferromagnetic thin-film traversed by a current in an orthonormal coordinate system of unit vectors x, y, and z, where x is the direction of current density J, and χ the angle formed in the plane of the film (x, y) between the magnetization and the current (Fig. 1). The points (1), (2), (3), and (4) are the locations of the measurement points, separated by a distance d. We can then measure a voltage V_{12} parallel to the current and a voltage V_{34} perpendicular to the current.

$$E_x = [\rho_{\perp} + \Delta\rho\cdot\cos^2(\chi)]\cdot J_x \qquad (1.5)$$

$$E_y = \Delta\rho\cdot\sin(\chi)\cdot\cos(\chi)\cdot J_x \qquad (1.6)$$

$$E_z = \rho_H\cdot\sin(\chi)\cdot J_x \qquad (1.7)$$

Based on Eq. 1.5, which is another way of representing Ohm's law, we can write:

$$\rho(\chi) = \rho_{\perp} + \Delta\rho\cdot\cos^2(\chi) \qquad (1.8)$$

This is the anisotropic magnetoresistance equation (see Chap. 2).

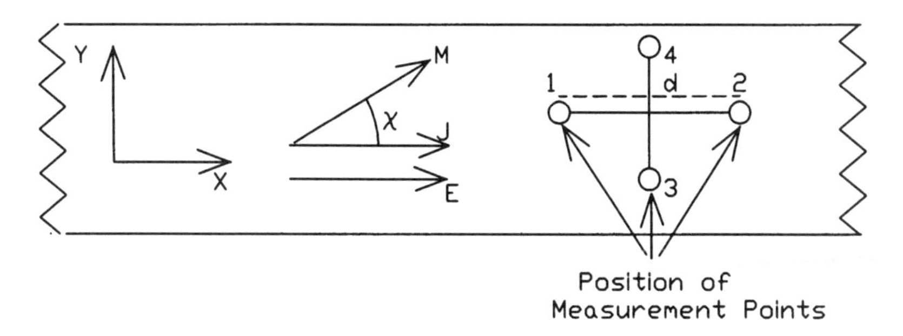

Fig. 1 Different vectors used for the phenomological equations of magnetoresistance in a layer of magnetoresistive material.

[3]Jan., *op. cit.*

Equation 1.6 shows that it is possible to measure a voltage perpendicular to the measured current and in the plane of the layer. This effect is sometimes referred to as the *planar Hall effect,* or *pseudo-Hall effect.* Its origin is identical to that of the magnetoresistance.[4]

Equation 1.7 represents the *extraordinary,* or *spontaneous, Hall effect,* which is negligible in the materials used in magnetoresistive sensors.

Based on these equations, we can define the following voltages:

$$V_{12} = J\rho d\left[1 + \frac{\Delta\rho}{\rho}\cdot\cos^2(\chi)\right] \qquad (1.9)^5$$

This represents the magnetoresistive signal, which is independent of the polarity of the angle $[V(\chi) = V(-\chi)]$. This equation also shows the weak dependence of the signal on the angle χ when the magnetization is aligned with the x or y axis. The optimum sensitivity is obtained for $\chi \cong 45°$. The voltage V_{34} along the y axis is equal to:

$$V_{34} = J\rho d\left(\frac{\Delta\rho}{\rho}\cdot\sin(\chi)\cdot\cos(\chi)\right) \qquad (1.10)^6$$

3. STRUCTURE AND OPERATING PRINCIPLES

3.1 Hysteresis Cycles

Ferromagnetic thin films are applied using vacuum deposition or sputter techniques. During deposition, a magnetic field is applied in the direction of the film. The presence of this field generally induces a uniaxial magnetic anisotropy along the plane of the deposited layer.

Magnetization in these layers is spontaneously oriented along an axis referred to as the *easy axis* (EA). The term refers to the fact that when no magnetic field is applied to the film, the magnetic moments present in the material tend to align themselves along this axis. Mathematically this situation can be represented by introducing an anisotropic energy density E_k in the general equation for free energy, so that $E_k = K_u \sin^2\varphi$ in which φ is the angle between the saturation magnetization moment and the applied magnetic field (Fig. 2).

[4]Fluitman, J. H. J. and Groenland, J. P. J., *IEEE Trans. Magn.* Mag 17, 6 (1981): 2893.

[5]In current magnetoresistive materials, $\Delta\rho$ is on the order of several percent. Consequently, $\rho_\perp$ can be replaced by ρ in many practical applications.

[6]See previous footnote, where $\rho_\perp \cong \rho$.

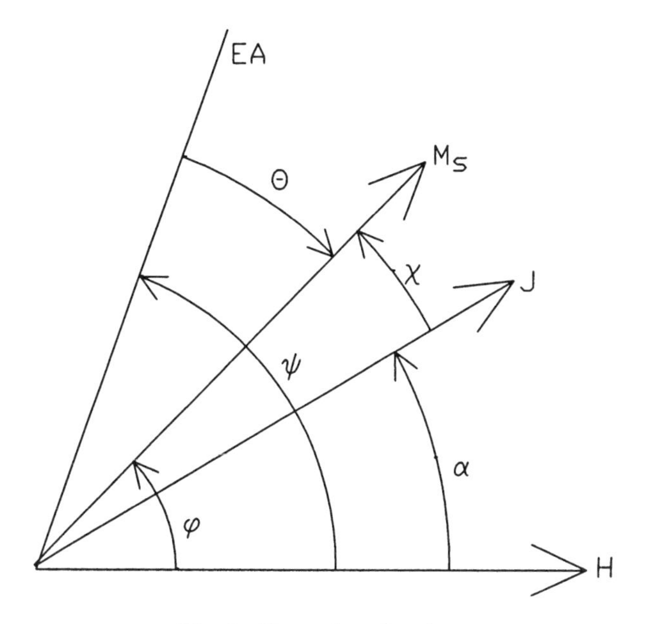

Fig. 2 Illustration of angles.

The magnetic behavior of a film displaying uniaxial anisotropy can be approached through the use of a model based on the coherent rotation of magnetization, or Stoner-Wohlfarth[7] model.

In a material displaying a constant uniaxial anisotropy K_u and a saturation magnetization M_s in the presence of a magnetic field H, the *magnetic energy density* is given by:

$$E = K_u \cdot \sin^2(\varphi - \psi) - M_s H \cdot \cos \varphi \tag{1.11}$$

in which ψ is the angle between the easy axis and the field.

The direction of the magnetization equilibrium, or the angle φ, is obtained by solving the system of equations:

$$\frac{\partial E}{\partial \varphi} = 0 \quad \text{and} \quad \frac{\partial^2 E}{\partial \varphi^2} \geq 0 \tag{1.12}$$

which determine the minimum energy for the system.

From this solution we can determine the shape of the hysteresis cycles for the film by calculating φ for the entire field H, and treating

[7]Chopra, K. L., *Thin Film Phenomena.* Chap. 10, 614. R. E. Krieger Publishing, 1979.

$\cos\varphi$ as a function of H. In particular, when $\psi = \pi/2$ (transverse cycle), Eq. 1.11 becomes:

$$E = K_u \cdot \cos^2\varphi - M_s H \cdot \cos\varphi \tag{1.13}$$

and the system of equations in Eq. 1.12 becomes:

$$\sin\varphi \cdot (M_s H - 2K_u \cdot \cos\varphi) = 0 \tag{1.14a}$$

$$\cos\varphi \cdot (M_s H - 2K_u \cdot \cos\varphi) + 2K_u \cdot \sin^2\varphi > 0 \tag{1.14b}$$

The solution for $\cos\phi$ is then:

$$\cos\varphi = -1 \quad \text{for} \quad H < -\frac{2K_u}{M_s}$$

$$\cos\varphi = \frac{M_s}{2K_u} \cdot H \quad \text{for} \quad -\frac{2K_u}{M_s} < H < \frac{2K_u}{M_s}$$

and

$$\cos\varphi = 1 \quad \text{for} \quad H > \frac{2K_u}{M_s}$$

The term $2K_u/M_s$ has the dimensions of a field $(L^{-1}t^{-1}Q)$ in A · turn/meter. We can then define $H_k = 2K_u/M_s$, which is the anisotropy field of the material. This field is needed for magnetization to switch reversibly between the easy and hard axes. Figure 3 represents the respective hysteresis cycles for $\psi = \pi/2$ and $\psi = 0$. When $\psi = \pi/2$, the cycle obtained shows neither hysteresis nor discontinuity, only a reversible rotation of the magnetization.

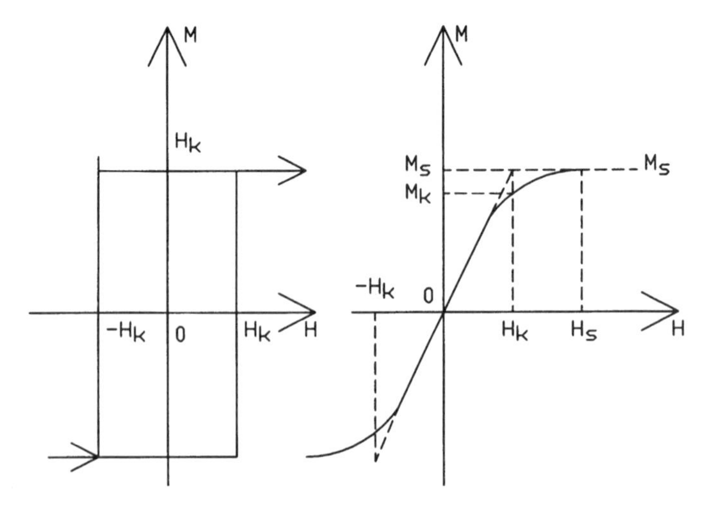

Fig. 3 Hysteresis cycles for $\psi = \pi/2$ and $\psi = 0$, respectively.

The magnetoresistive response of a thin layer is, therefore, determined by the relative dependence of the magnetic moment and its direction with respect to the applied field. When a current passes through a thin layer, we can calculate its resistivity as a function of the angle φ and α, where α is the angle between the current and the applied field. We then have the equation $\chi = \varphi - \alpha$ and Eq. 1.8 becomes:

$$\rho(\varphi) = \rho + \Delta\rho \cdot \cos^2(\varphi - \alpha) \tag{1.15}$$

The most useful case for practical applications occurs when $\psi = \pi/2$. Equation 1.15 can then be written:

$$\rho(\varphi) = \rho + \Delta\rho \cdot (1 - \cos^2 \varphi) \tag{1.16}$$

Assume we have a flat film with a high L/h ratio (L being thc length, h the width), through which a constant current flows (Fig. 4). The easy axis is parallel to L and the current density. The film is thus divided into antiparallel magnetic domains aligned along the easy axis. The moments turn by an angle φ when a magnetic induction $B_Y = \mu_0 H_Y$ is applied to the film plane. By ignoring demagnetization effects, we can write:

$$\frac{M_Y}{M_S} = \sin(\varphi) = \frac{H_Y}{H_S} \quad \text{for} \quad -H_K \leq H_Y \leq H_K \tag{1.17}$$

in which M_Y and H_Y are, respectively, the component of magnetic moment in the direction of the y axis and the component of the field H in the same direction.

By introducing this in Eq. 1.16, we obtain:

$$\rho(\varphi) = \rho_0 + \Delta\rho[1 - (H_Y/H_K)^2] \tag{1.18}$$

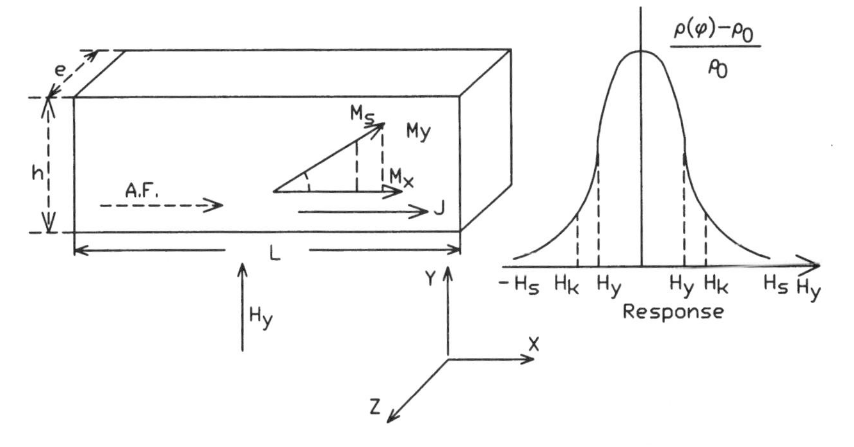

Fig. 4 Response of a sensor when edge effects are not negligible.

The magnetoresistive response of a sensor constructed in accordance with these principles can therefore be written as:

$$\Re = \frac{\rho - \rho_0}{\rho_0} = \frac{\Delta\rho}{\rho_0}\left[1 - (H_Y/H_K)^2\right] \tag{1.19}$$

The resistance of the sensor is at a maximum when H_Y is zero and decreases as H_Y increases, so that:

$$\Delta R/R_{\max} = \cos^2(\varphi) \tag{1.20}$$

in which $R_{\max}$ is the value of the resistance when the magnetization M is parallel to the current density. The direction of H_Y does not affect the value of $\Delta R\left[V(\varphi) = V(-\varphi)\right]$. Figure 4 shows the response of a sensor as a function of H_Y. The dotted line corresponds to the ideal quadratic response (Eq. 1.15), in which the demagnetizing field has been ignored. Edge effects resulting from this field produce the solid curve shown in the illustration. Signal saturation occurs when the value of the applied field exceeds the value of the demagnetizing field H_D $(H_S = H_K + H_D)$, which is an anisotropic field that can be expressed as $H_D = (e/h) \cdot M_S$.

Another cause of response deterioration compared to what we would expect from the equation in $\cos^2(\varphi)$ is the angular dispersion of the easy axis. In this case the angle φ must be replaced with $\varphi \pm \alpha$ and we see that the same applied field can have two resistance values. This dispersion can result from mechanical or thermal strain within the thin film. Figure 5 illustrates the effect of such strain on the magnetoresistive response.

If we consider the magnetoresistive element shown in Fig. 4, for $H_Y = 0$, we have $V(H = 0) = \rho_{\|} \cdot I \cdot L/(h \cdot e)$, $(\varphi = 0)$ and for $H_Y = H_S$, we have $V(H = H_s) - (\rho_{\|} - \rho_{\perp}) \cdot I \cdot L/h \cdot e$, $(\varphi = \pi/2)$.

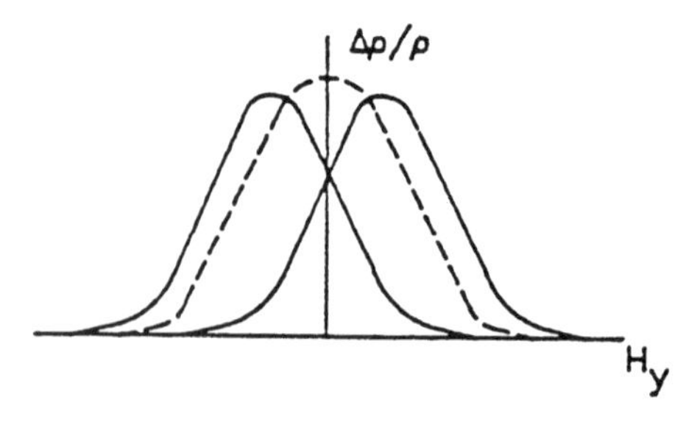

Fig. 5 Effects of dispersion on magnetoresistive response. Two signals are superimposed, introducing mechanical or thermal strains.

The difference between these voltages results in the maximum magnetoresistive signal supplied by the device. This signal is proportional to the anisotropy of resistivity, $\Delta\rho$, which is obviously one of the criteria used in material selection. For this reason, the ratio $\Delta\rho/\rho$ must be as high as possible.

We also see that, according to Eq. 1.19, the signal is proportional to the field crossing the magnetoresistive element and *not the derivative of the flux with respect to time,* as it is in an inductive sensor. *This characteristic is very useful when working with encoders or position sensors.* The rotational speed of a rotary or position sensor does not affect the magnitude of the signal, which is only affected by the value of the magnetic field H.

4. MATERIAL SELECTION

Binary alloys. Of the nickel-based ferromagnetic alloys, the most magnetoresistive is $Ni_{70}Co_{30}$, with a $\Delta\rho/\rho = 3.8\%$ for $e = 300\text{Å}$. The resistivity of the alloy is low since Ni and Co possess a shared d band. Its magnetostriction and high coercive force ($H_c = 119$ AT/m) make it somewhat unreliable in actual use, however. The $Ni_{81}Fe_{19}$ alloy, for example, has a magnetostriction coefficient of zero.

Ternary alloys. There is a number of NeFeCo alloys that are non-magnetostrictive. These alloys also display greater magnetoresistance than NiFe alloys and better response. Table 1 lists some of the properties of these alloys.

TABLE 1[8] Properties of NiFeCo Alloys

Composition Ni:Fe:Co	ρ_0 $\mu\Omega\cdot$cm	$\Delta\rho/\rho_0$ %	H_c ATm^{-1}	H_k ATm^{-1}	α deg
82:18:00	40.00	1.30	120	265	0.30
74:16:10	33.50	1.65	150	1025	0.15
65:15:20	25.00	2.20	135	1665	0.10
60:10:30	20.00	2.60	150	1930	0.05
55:05:40	41.50	1.13	230	2180	0.20

[8]Collins, A. J. and Sanders, I. L., *Thin Solid Films.* 48 (1978): 247.

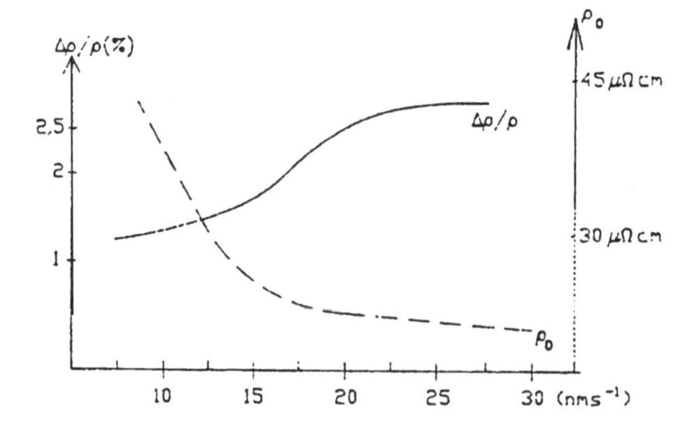

Fig. 6 Experimental variations of $\Delta\rho/\rho$ and ρ_0 vs. deposition rate of NiFeCo (in $\mathrm{nm} \cdot \mathrm{s}^{-1}$).

The 65:15:20 alloy in the table above combines high magnetoresistance with a magnetostriction that is nearly zero ($\lambda_S \cong 4 \cdot 10^{-7}$).[9]

Deposition conditions also have a certain influence on the magnetic properties of the alloy. Figure 6 shows the variation in $\Delta\rho/\rho_0$ and ρ_0 as a function of deposition rate. Imperfections in the crystal lattice, which are inherent in thin films, contribute to the difference in resistivities and magnetostriction coefficients between thin films and solids. To reduce the value of α, it is always necessary to vacuum-anneal these films in the presence of a magnetic field.

In practice, ternary alloys are preferred to NiCo alloys because of their low coercive force and magnetoresistive response. The first choice among the ternary alloys is $Ni_{65}Fe_{15}Co_{20}$.

5. SENSOR DESIGN

In practice, sensors are made of a magnetoresistive thin film deposited on an insulating substrate, either glass or oxidized silicon. After annealing in a vacuum and applying a magnetic field to reduce the angle α, the thin film is etched using standard methods (chemical, plasma, etc.). Gold conductors are then deposited and etched to direct current through the sensor

[9]λ_S is estimated by measuring the hysteresis cycles by means of the Kerr effect under controlled conditions and in the presence of a magnetic field perpendicular to the easy axis. When polarized light passes through a magnetic film, the plane of polarization turns by an angle of $1/3°$, which is proportional to the component of the field parallel to the direction of the beam.

and measure the voltage drop at the resistor terminals. Through the use of microelectronics technology, integrated sensors designed for specific applications[10] have been developed.

Typical sensor lengths, which correspond to track widths in magnetic recording media, are given below:

- $2 \times 610\ \mu m$ standard analog cassette
- $250\ \mu m$ floppy disk (96 tracks per inch)
- $100\ \mu m$ hard disk (254 tracks per inch)
- $50\ \mu m$ hard disk (508 tracks per inch)
- $26\ \mu m$ hard disk (1,000 tracks per inch)

The width varies with the application between $20 \mu m$ and $10 \mu m$. The standard thickness of the films is roughly 5000nm. In the case of angular and longitudinal encoders, L will be approximately $50 \mu m$. Figure 7 is an illustration of a typical sensor head with its active thin film in NiFeCo sputtered on a silicon dioxide layer and its gold conductors. Table 2 lists some of the characteristics of these sensors.

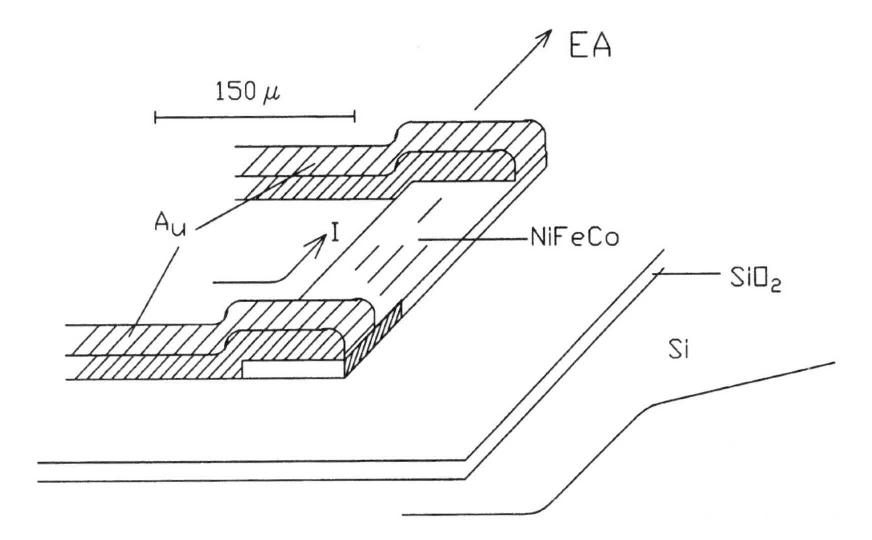

Fig. 7 Schematic representation of a sensor head on a silicon substrate.

[10]Methods exist to linearize the signal (magnetic bias, barber poles, etc.) used in sensors for audio or video recording. For lack of space we will not discuss these methods here. Readers interested in the topic should consult P. Ciureanu and H. Gavrila, "Magnetic Heads for Digital Recording." *Studies in Electrical and Electronic Engineering*. Elsevier, 39 (1990).

TABLE 2 Characteristics of Magnetoresistive Sensors

	NiFe	NiFe	NiFeCo	NiFeCo
L (μm)	610	610	610	610
h (μm)	12	6	14	8
e (angstrom)	380	380	310	310
H_k (ATm^{-1})	557	557	1114	1114
M_S (Wm^2)	10	10	12	12
H_S (ATm^{-1})	3103	5570	3382	4974
ΔV_{max} (mV)	30	30	32	33
J (A/cm^2)	$8.8 \cdot 10^5$	$9.6 \cdot 10^5$	$8.5 \cdot 10^5$	$8.9 \cdot 10^5$
$\Delta\rho$ ($\mu\Omega \cdot$ cm)	0.56	0.51	0.62	0.61

6. APPLICATIONS[11]

6.1 Incremental Magnetic Encoders

Magnetoresistive sensors can be used as sensing devices in incremental encoders and tachometers. The principle of operation is illustrated in Fig. 8. Depending on the application, we can measure an angle or angular velocity. These devices have the following advantages compared to optical technologies:

1. Good stability at high rotational speeds (3,000 to 8,000 rpm)
2. Good resistance to humidity and vibration (automotive use)
3. Very little dissipation
4. Operation and reliability are only slightly dependent on temperature
5. Fast response

The surface of a drum or disk is covered with a thin- or thick-film magnetic material (particles of γFe_2O_3 or CrO_3 having a coercive force of 24,000–31,000 ATm^{-1}).

[11]Miyashita, K., et al., "Absolute Magnetic Rotary Encoder." *IEEE IECON,* 1985.

Campbell, P., "A Low-Cost Positioning Magnetic Encoder." *Proceedings Motor-Con.* 1987. INSPEC Conference Paper, September 1987.

Sei, M., et al., "Durable and Multifunctional Magnetic Rotation Sensor for Rotary Encoder." *IEEE Trans. Mag.* Mag.-23. 5 (1987): 2185.

Yamashita, S., et al., "Development of the Absolute Magnetic Encoder Suitable for Thinner Structure." *IEEE Trans. Mag. Japan.* 5, n.8, (August 1990): 711.

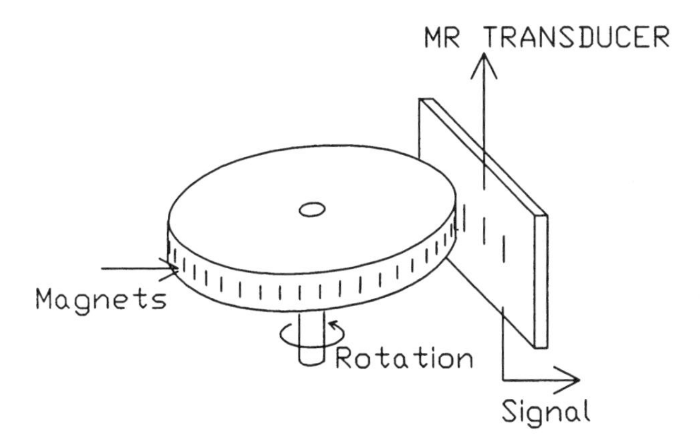

Fig. 8 Mechanical principle of a simple magnetoresistive absolute encoder.

This medium is then magnetized until saturation, using an AC signal having a pitch P lying between two magnetic transition domains. The MR sensor consists of several resistive bands separated by a distance $P/2$. The horizontal flux between domains penetrates the sensing elements in a direction perpendicular to the easy axis. An AC signal is then obtained at the output of a measuring bridge, the periodicity of this signal corresponding to the pitch P and the number of transitions N corresponding to the angle $N \cdot P/(2\pi R)$. The number of cycles in a 360° rotation of the disk represents the resolution of the system. To maximize the signal the distance between the sensor and the magnetic medium should be $0.7P$. For example, a sensor having a 50mm disk with a pitch of 60μm has a resolution of 2,500.

6.2 Absolute Magnetic Encoders

Absolute encoders work in exactly the same manner as incremental encoders. The difference lies in the use of several tracks, which generate signals in response to a specific code (Gray code, for example). Figure 9

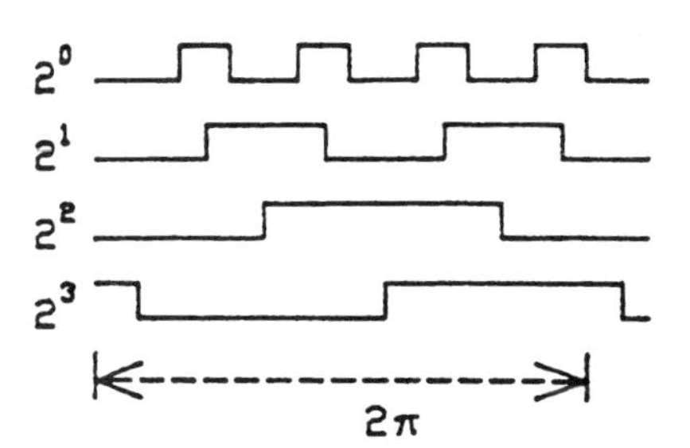

Fig. 9 Ideal response of a 4-bit absolute encoder.

illustrates the ideal output for an absolute 4-bit encoder. However, because of the response of the magnetoresistive material to the magnetic field (see Fig. 4), in practice the output signals will have the shape represented in Fig. 10.

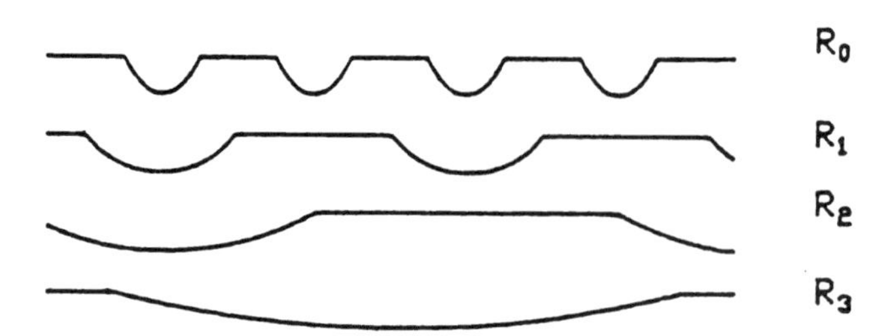

Fig. 10 Real response of the same encoder. The degradation is due to edge effects.

Specialized literature describes several methods for improving the response of the device. One example, developed by Hitachi, is shown in Fig. 11.

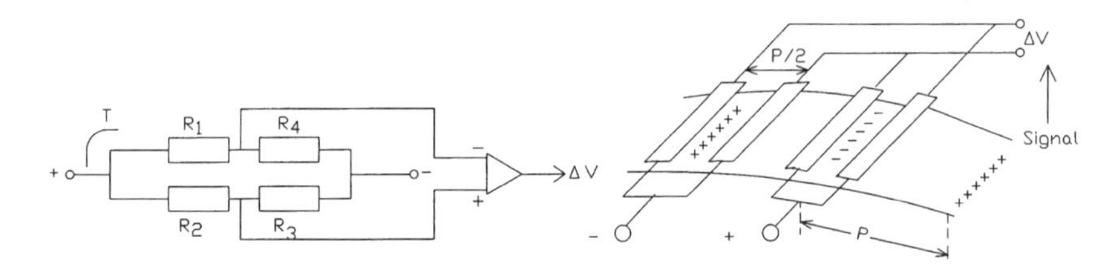

Fig. 11 Electronic circuit used to improve the response. Schematic of measuring bridge and physical implementation of NiFeCo resistive elements with their connections.

The magnetic track, consisting of four elements, contains parallel magnetic poles of identical width but different pitch, corresponding to the different bits. The sensor consists of four resistive elements, parallel to the tracks. Because the slope of the change in resistance depends on the length *h* of each sensor, the response is improved. To increase the number of bits, however, the pitch of each pole must decrease, for a given sensor diameter, in proportion to the number of bits. The intensity of the field will then decrease; hence, maintaining a noise-free signal will be difficult. This type of configuration can be used in low-resolution encoders such as pole detectors in electric motors. Figure 12 shows the improved response of the sensor.

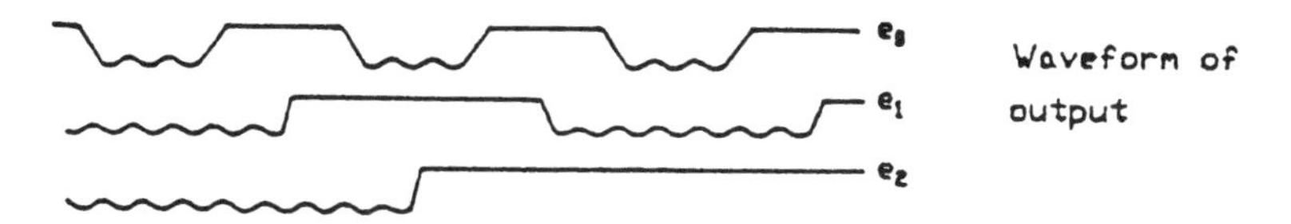

Fig. 12 Actual response of the encoder using Fig. 11 circuit.

Resolution can be improved, however, through the use of new recording materials that can take advantage of magnetic moments perpendicular to the track. Under these conditions the pitch can be reduced in size without any field loss. An improved detection method based on this principle has been developed in Japan by Hitachi. Readers interested in these methods will benefit from reading the references quoted in footnote 11, in particular the reference to Hitachi which explains the design of a 12-bit encoder. Its characteristics are shown below.

Resolution (bits)	4096
Frequency (kHz)	200
Output voltage (V)	5
Operating temperature (°C)	−10 to 80
Diameter (mm)	50

Even greater resolution could be obtained if the magnetic moments are recorded perpendicular to the track. This method could provide 16 bits without signal loss.

7. CONCLUSION

Magnetoresistance, which was a laboratory curiosity thirty years ago, is now used to develop hard drives for computers and high-resolution position sensors. The materials currently in use have the disadvantage of their low magnetoresistance coefficient, on the order of a few percent of the principal resistance. This currently limits their use in professional applications, which contain additional circuits that are free from noise. New materials with "giant" magnetoresistance coefficients (see Chap. 2) are now being studied in laboratories around the world.[12] Their use in consumer products, audio and video recording in particular, is only a few years away.

[12]Sensors based on giant magnetoresistance (called GMR) permit storage up to 0,2 to 2 Gbytes/cm^2. However, these sensors, which are constituted by about 60 thin-film layers of alternatively ferromagnetic and nonmagnetic materials, are difficult to manufacture. Recently, MIT laboratories announced a new type of sensor called junction magnetoresistive resistance (JMR). Two thin films of magnetoresistive material are separated by an ultra-thin film of magnesium or aluminum oxides (between 1nm and 2nm thick). When a voltage is applied between the two magnetoresistive layers, an electric current circulates through the oxide layer by quantum tunnel effect. Since the oxide layer is an insulator, its ohmic resistance is very large, and relatively large voltage signals are produced. When the JMR sensor detects a very weak change in magnetic field, it transmits a potential of the order of tens of mV. These sensors are therefore more sensitive than the GMR, and it is estimated that they will contribute to increasing by several times the present Gbytes density.

Magnetic storage is however not the only application. Development laboratories are presently working on sensors such as sensitive detectors for shaft speed, automotive antilock brakes, magnetic switches, magnetic accelerometers—and more generally, on applications requiring fast response, low power and low cost. Interested readers should consult http://www.ornl.gov/ORNLReview.

13

Nonlinear Resistors

1. INTRODUCTION

In the previous chapters we examined resistive components whose mechanical and electrical properties must remain stable during operation. To avoid degrading the performance of the electronic circuits in which they are used, it is important that changes in nominal resistance due to thermal, electrical, mechanical, and chemical factors, as well as humidity, be kept to a minimum. The techniques and materials used to satisfy this condition have already been discussed.

The introduction of signal measurement and processing technologies resulted in a demand for new components designed to measure and interact with physical parameters such as temperature and pressure, the concentration and composition of gas mixtures, and changes in voltage. A large number of these components are based on measurements of changes in resistance as a function of these parameters. To achieve the highest possible sensitivity, researchers have continued to develop materials in which conductivity *variation* as a function of the magnitude of the physical phenomena being measured is maximized. This approach is therefore antithetical to that used in the development of conventional resistors. In addition, the materials used must also satisfy certain criteria of selectivity. A component intended to measure the concentration of carbon dioxide gas should not be sensitive to any phenomena other than those being measured.

This chapter describes such components, which are referred to as nonlinear resistive sensors. For each type of sensor, the following elements will be described and analyzed: properties of the materials used, fabrication technologies, and applications. Among the many kinds of resistive sensors on the market, we will concentrate on negative temperature coefficient (NTC) and positive temperature coefficient (PTC) thermistors, varistors, and oxygen gas sensors.

2. CONDUCTIVITY MECHANISMS IN CERAMICS AND OXIDES

Nonlinear resistors are based exclusively on the use of metal oxide ceramics.[1] Their characteristics must be suitable for the manufacture of sensors. Such ceramics have semiconductor properties, which we will examine briefly. However, while the transport of free charges (electrons and holes) within the grains determines the characteristics of NTC sensors, interface phenomena at the grain boundaries determine the overall electrical properties of PTC resistors and varistors. The properties of gas resistance sensors are governed by the exchange of chemical species between the solid and its environment, resulting in nonstoichiometric states within the crystal, which lead to conductivity variations in certain ceramics.

2.1 Electrical Conductivity of Ceramics

As shown in Chap. 1, the electrical conductivity in a solid is equal to the sum of the conductivity of the different carriers

$$\sigma = \sum_j \sigma_j$$

Each partial conductivity, whose origins may be electronic or ionic, is in turn defined by $\sigma_j = N_j Z_j q \mu_j$, where N_j is the carrier density (cm^{-3}) of the

[1]There are a number of excellent references on the insulating, semiconductor, and conductor ceramics used in electronic components. These include:

J. B. Goodenough, *Oxydes des métaux de transition*. Paris: Gauthier-Villars, 1973.

R. C. Buchanan, ed., *Ceramic Materials for Electronics*. NY: Marcel Dekker, Inc. 1991.

M. F. Yan and A. H. Heuer, *Additives and Interfaces in Electronic Ceramics*. American Chemical Society, 1982.

K. L. Frober, *Ceramic Technology for Electronics*. ISHM series 6984-004, 1984.

species j, $z_j q$ the effective charge in coulombs (Cb), and μ_j the mobility (cm^2/V-sec) of the species. In the ceramics used in NTC thermistors, the mobility μ will be the principal factor, together with N_j, governing the change in conductivity with temperature. The ceramics used in nonlinear resistors will of course have semiconductor properties. Depending on the temperature, various mechanisms will affect this conductivity: intrinsic conductivity through the creation of electron-hole pairs in the conducting and valence bands and also through the creation of Schottky and Frenkel defects, extrinsic conductivity resulting from impurities in the crystal, *polaron* conductivity, and tunnel effects at grain boundaries. We will briefly describe these mechanisms below. It should be pointed out that among these conduction phenomena, polarons and grain boundary effects are most responsible for high absolute values of TCR and VCR (voltage coefficient of resistivity).

Intrinsic conductivity. The number of electron-hole pairs created on each side of the band gap of a semiconductor through thermal effects is given by the well-known formula:

$$np = AT^3 \exp\left[-\frac{E_g}{kT}\right] \tag{1.1}$$

Since each electron is compensated for by a hole:

$$n_i = p_i = A^{1/2}T^{3/2} \exp\left[-\frac{E_g}{2kT}\right] \tag{1.2}$$

A is a constant that depends on the effective mass of the carriers, and E_g is the value of the band gap. A gap of 1eV (Si) yields an n and p carrier density on the order of $10^{12}cm^{-3}$ at 500°K (compared to $5 \cdot 10^{22}cm^{-3}$ in metals). The gap varies between 1eV and 10–15eV (SiO_2), with intrinsic conductivity alone producing a sufficiently wide range of resistances and sensitivity to temperature. The creation of defects in ionic solids also serves as a mechanism for conductivity. It has been demonstrated that if N_i is the number of vacancy-interstitial pairs in the crystal lattice (see Chap. 3 for an explanation of defect creation phenomena), and ΔH_s and ΔH_f are, respectively, the enthalpy of the anion-cation vacancy (Schottky defect) and the interstitial ion vacancy (Frenkel defect) then:

$$[V_{m'}] = [V_{o'}] = N \exp\left[-\frac{\Delta H_s}{2kT}\right] \tag{1.3}$$

for Schottky defects and

$$[V_{m'}] = [M_i^*] = (NN_i)^{1/2} \exp\left[-\frac{\Delta H_f}{2kT}\right] \tag{1.4}$$

for Frenkel defects.[2] Here $[V'_m]$ represents the cation vacancy density, $[V'_o]$ the anion vacancy density, and $[M_i^*]$ the interstitial anion density. Since ΔH_s and ΔH_f are approximately 5–12 eV, NTC thermistors constructed with these materials will have an extremely high negative TCR.

Extrinsic conductivity. A certain number of impurities can lead to the formation of holes or electrons, depending on whether the levels are acceptors or donors. The number of donors and acceptors is given by the following equations:

$$n = BT^{3/4} \exp\left[-\frac{E_d}{2kT}\right] \quad \text{and} \quad p = CT^{3/4} \exp\left[-\frac{E_a}{2kT}\right] \tag{1.5}$$

in which E_d and E_a are the energies of ionization of the donor and acceptor levels, respectively, and B and C are constants proportional to the square root of the donor and acceptor concentration. Since the values of E_d and E_a are lower than the energy E_g or the enthalpies ΔH_s and ΔH_f, changes in temperature have almost no effect on the contribution to the total extrinsic conductivity of the component, at least within the range of operating temperatures of nonlinear resistors. One well-known example is the replacement of a Li^+ ion by a Ni^{3+} ion in NiO, which can be treated as a hole in the valence band.

Phase transformations. Sudden changes in conductivity occur in the neighborhood of temperatures that cause phase changes in the oxide. The change can be the result of a change in transport properties within the crystal or of phenomena occurring at the grain boundaries. These phenomena can be exploited in the nonlinear components used for circuit protection, such as $BaTiO_3$-based PTC devices.

Variation in mobility as a function of temperature. In nonpolar semiconductors the variation in mobility with temperature is limited by carrier-phonon interactions (see Chap. 3) and assumes the form:

$$\mu = \mu_o(m^*)^{5/2^* T^{-3/2}} \tag{1.6}$$

The orders of magnitude of μ in these semiconductors is approximately 600 (ZnTe) to 70,000 cm^2/V-sec (InSb) at 300°K.

In ionic oxide compounds, on the other hand, carrier mobility is largely determined by the interaction with polarized ions within the crystal. The carrier (electron) causes a deformation of the lattice, and through polarization, produces a potential well that restricts its movement to the

[2]F. A. Krogel, *The Chemistry of Imperfect Crystals.* North-Holland Amsterdam, 1974.

vicinity of the ion. The electron-polarizing field pair is known as a *polaron.* When the radius of the potential well reaches several interionic distances, the electron continues to move in the conduction band but with a greater effective mass. In this case it is shown that:

$$\mu = \mu_o T^{-1/2} \tag{1.7}$$

The value of μ is between 1 and 150 cm^2/V-sec in these materials. However, if the dimension of the well is approximately one interionic distance, the electron will remain at the site (small polaron) and will only be able to move to an adjacent site by hopping to it. Under these conditions:

$$\mu = \frac{\mu_o}{kT} \exp\left[-\frac{E_h}{kT}\right] \tag{1.8}$$

in which E_h is the hopping energy. Since the majority of ceramics used in nonlinear resistors consist of transition metal oxides, the very narrow 3d band contributes to the creation of a large number of small polarons that contribute to conductivity phenomena. For most oxides the value of E_h is approximately 0.1–0.4eV.[3]

Influence of grain boundaries on conductivity. Because of their nature, grain boundaries represent inhomogeneities in the structure and composition of the material. Such inhomogeneities of composition are produced whenever impurities migrate throughout the crystal and attach themselves to the grain boundaries. In doing so they also introduce crystal structure inhomogeneities. These defects and impurities, generally polarized, give rise to potential barriers ϕ_{BO} and space charges. We can use Poisson's equation to calculate the value of this potential. The equation takes the form:[4]

$$\phi_{BO} = \frac{qN_b^2}{8\varepsilon\varepsilon_o N_d} \tag{1.9}$$

in which N_b is the charge density at the interface of width d, N_d is the donor density within the body of the crystal, and ε is the relative dielectric constant. This potential barrier plays a fundamental role in conductivity since the probability of electrons or holes crossing it is proportional to $\exp(-\phi_{BO}/kT)$. Figure 1 provides energy and spatial representations of the potential barrier.

[3]J. B. Goodenough, *Les oxydes des métaux de transition.* Paris: Gauthier-Villars, 1973.
[4]W. G. Morris and I. W. Cahn, *Grain Boundaries.* Baton Rouge: Claitors, 1975.

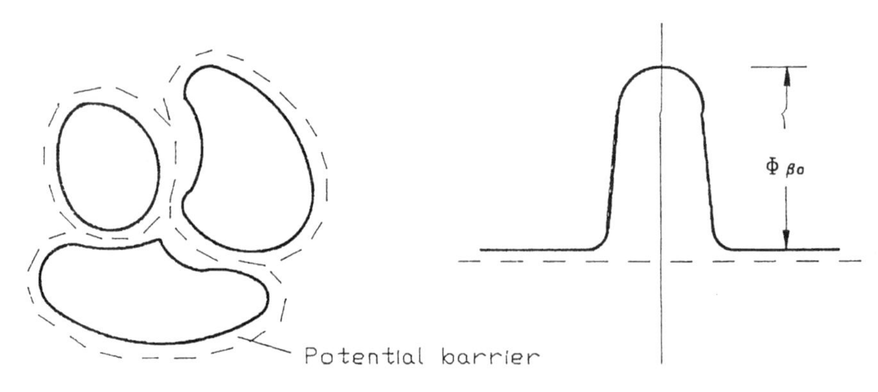

Fig. 1 Spatial and energy representation of the potential barrier.

2.2 Ionic Conductivity in Ceramics

Variations in conductivity in ionic ceramics are due to changes in the stoichiometry caused by redox reactions. These in turn are caused by the gas or liquid species E_i, whose concentration we wish to measure and which diffuses to the interior of the crystal from its surface. This results in the creation or annihilation of donors or acceptors in the crystal. Under these conditions it has been shown that the variation in conductivity follows a law of the type $\sigma \cong p^{-n}$, in which p is the partial pressure of the gas being analyzed (or the concentration in case of a liquid) and n a coefficient between 0 and 1.

For example, titanium dioxide, TiO_2, is typically used as an oxygen detector. The increased conductivity in a reducing atmosphere results from a loss of oxygen by the oxide, which passes from an oxidized (TiO_2) to a reduced state (TiO_{2-x}). The coefficient x characterizes the deviation from stoichiometry.

From a chemical point of view—and in order for a state of equilibrium to exist for each value of x—this results in the creation of atomic defects represented by oxygen vacancies $[V_c]$ and the presence of interstitial titanium [Ti]. When the oxygen vacancies are present, the equilibrium between the oxygen of the crystal and the gas can be expressed as:

$$O_c^x \Leftrightarrow \frac{1}{2} O_{2(G)} + [V_c^x] \tag{1.10}$$

The vacancy created is subject to ionization according to the following:

$$[V_c^x] = [V_c'] + e \quad \text{and} \quad [V_c^x] = [V_c''] + 2e \tag{1.11}$$

Consequently, $\sigma = [e]\mu_e$ shows that the conductivity is directly related to the extent of reduction of the crystal, and thus to the partial pressure of oxygen. Figure 2 represents the conductivity curve of TiO_2 as a function of the partial pressure of oxygen.

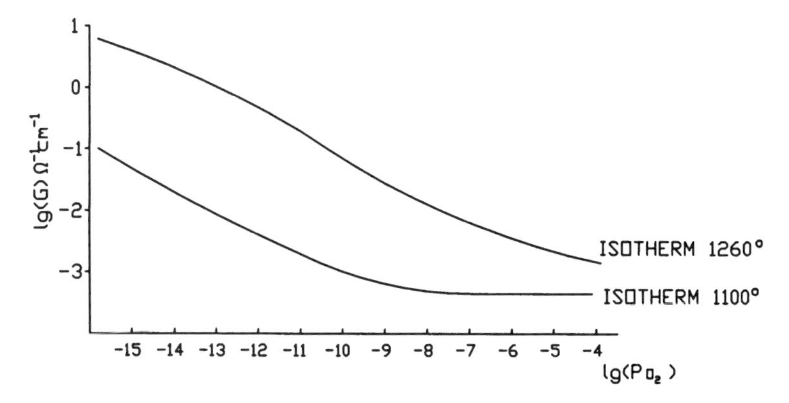

Fig. 2 Conductivity of TiO_2 as a function of the O_2 partial pressure.

Table 1 lists the different ceramics used to manufacture nonlinear resistors.

TABLE 1 Ceramics for Nonlinear Resistors

Component	Ceramic material
NTC thermistors	Transition metal spinels
PTC thermistors	Barium titanate ($BaTiO_3$)
Varistors	ZnO
Chemical sensors	
• gas	TiO_2, ZrO_2, SnO_2, etc.
• liquid	Al_2O_3, $MgCr_2O_4$

The choice of ceramic is dictated by its physical properties as well as the following parameters:

- Ease of manufacturing
- Constant chemical composition from batch to batch
- Constant electrical response over time
- Thermophysical properties
- Good mechanical properties
- Performance (stability, reliability as a function of time *t* and temperature T °K)
- Environmental constraints

3. NEGATIVE TEMPERATURE COEFFICIENT THERMISTORS

NTC thermistors are manufactured from semiconducting ceramics consisting of a blend of metal oxides sintered at high temperatures and generally crystallizing as a spinel structure.

Spinel structure oxides follow the chemical formula AB_3O_4. The structure consists of a compact cubic stack of anions (FCC) in which 16 octahedral B sites and 8 tetrahedral A sites are occupied per unit cell. Each B ion has 6 identical neighbors at a distance of $R = a_o(2)^{1/2}/4$. The spinel structure results in the formation of d bands extended throughout the B cation sublattice, allowing the formation of polarons. Figure 3 represents the unit cell of the spinel structure.

As indicated previously, these ceramics are characterized by a significant variation in conductivity with temperature.

$$\sigma = N_i \mu q = \sigma_o \exp\left[-\frac{E_h}{kT}\right] \tag{1.12}$$

Equation 1.12 is identical to Eq. 1.8.

The ratio E_h/kT corresponds to the sensitivity index B for a given thermistor. Its value is between 2000 and 6000°K and is related to the temperature coefficient, or TCR, by:

$$\text{TCR} = \frac{-B}{T^2} \tag{1.13}$$

The values of σ and B can be modified by changing the ceramic's chemical formulation; that is, by modifying the concentration and nature of the metallic oxides. Figure 4 represents such a variation.

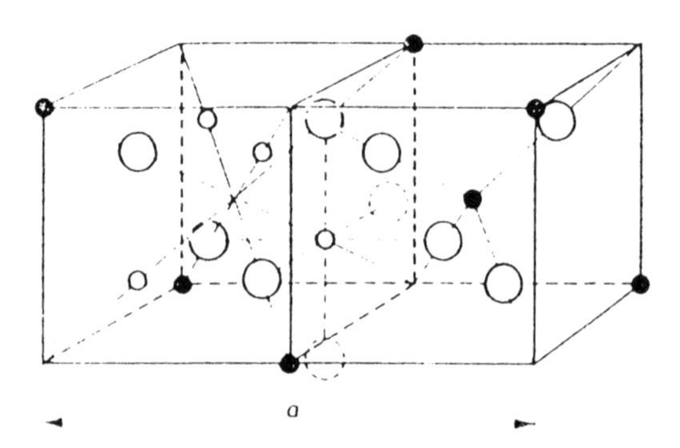

Fig. 3 Unit cell of the spinel crystal structure.

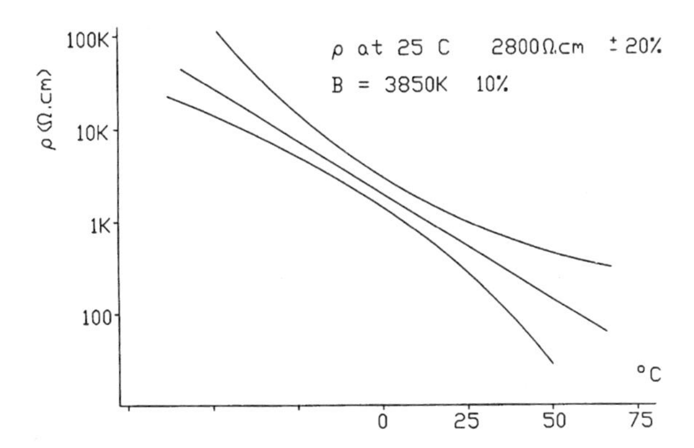

Fig. 4 Mn-Ni-O compound: resistance vs. temperature-average and dispersion.

For example, using the Mn-Ni-O diagram leads to the values shown in Fig. 5.

The decrease in resistivity with the concentration of Ni^{2+} can be explained by the mechanism of valence induction. To balance the system's electrical charges, each Ni^{2+} ion induces a 4+ valence in the manganese ion. An exchange of electrons then takes place between the Mn^{3+} and Mn^{4+} ions located at the octahedral sites of the spinel structure. The greater the concentration of Ni^{2+}, the greater the concentration of Mn^{4+}, thus promoting electrical conductivity. The minimum theoretical resistivity ($x = 0.66$) is in good agreement with the observed experimental value shown in Fig. 5 and corresponds to equal concentrations of Mn^{3+} and Mn^{4+}.

Other chemical diagrams are used to accommodate the different values of resistivity and TCR needed for specific applications. The most

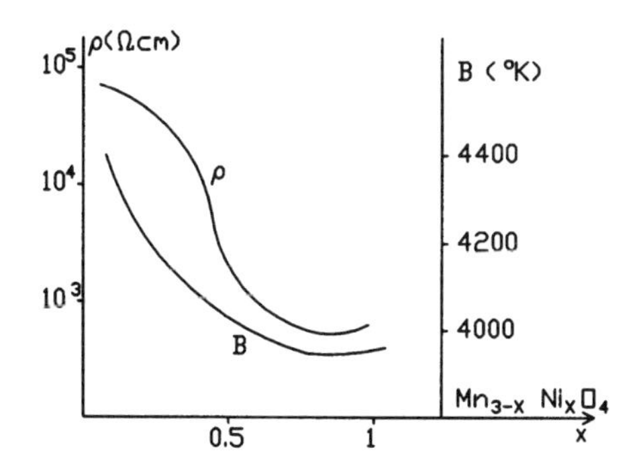

Fig. 5 Variation of B index and of resistivity vs. concentration of metallic oxides.

commonly used diagrams are: Mn-Co-O, Mn-Ni-Co-O, Mn-Ni-Ti-O, Mn-Ni-Cu-O, etc.

Figure 6 represents the variation in resistivity as a function of temperature for several NTC thermistors used in industrial and consumer applications.

The curves can be represented by a polynomial of the following type:

$$R_T = \left[\frac{\alpha}{T} + \frac{\beta}{T^2} + \frac{\gamma}{T^3} + \cdots \right] \tag{1.14}$$

The tolerance of ρ or R_T and B is generally between ± 1–10% for ρ and -4–6% for *standard* thermistors. Manufacturers can, upon request, supply lower tolerances, depending on the application.

For most applications the stability of the thermistor's characteristics with time and temperature is critical. Figure 7 represents the relative change in resistance over time at a temperature of 125°C for different compositions. There are two possible reasons for the increase in resistance:

1. The type of ceramic used: Ions are reorganized throughout various sites in the crystal and there is an exchange of oxygen with the atmosphere.
2. Changes in the ceramic-metal contact in electrodes.

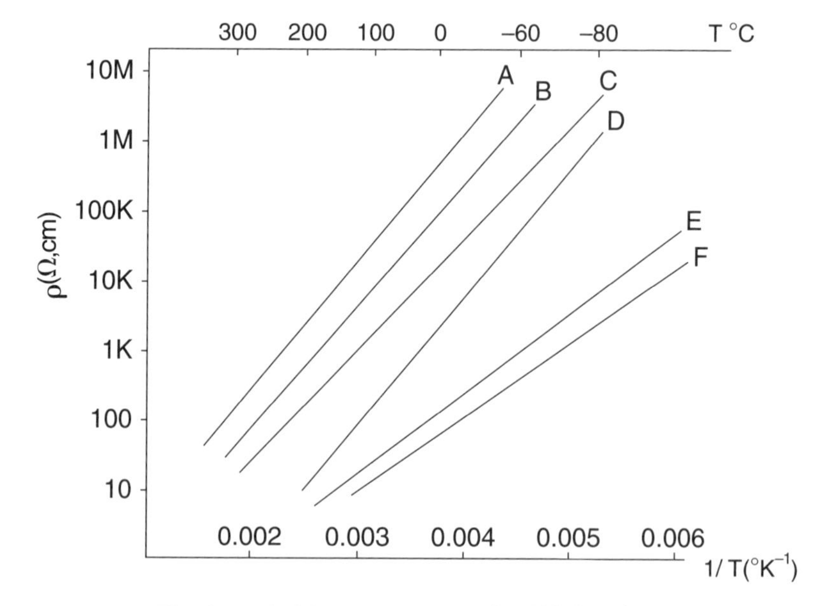

Fig. 6 Resistivity vs. temperature in NTC thermistors.

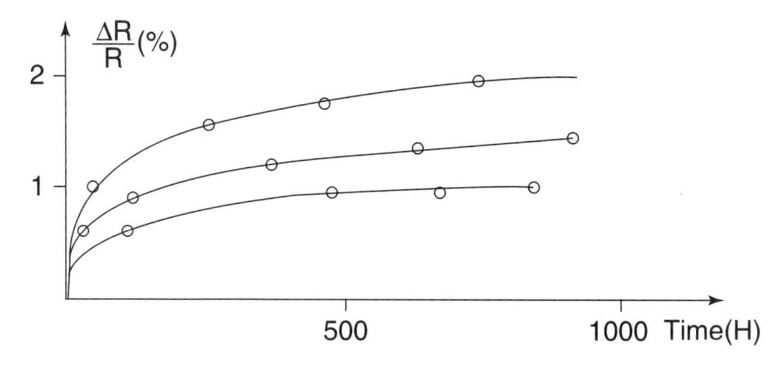

Fig. 7 Resistance drift with time at 125°C.

Generally, (1) is the most significant factor, and drift is related to the initial value of the material's resistivity, as it is in conventional resistors.

3.1 Production Techniques for Ceramic and for Electrode Deposition

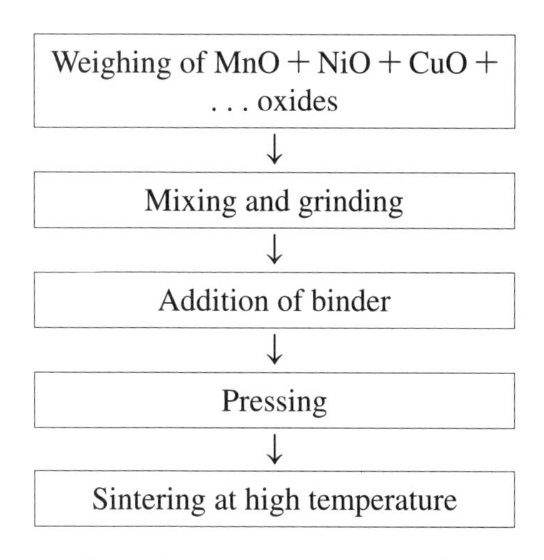

Production steps of thermistor ceramic.

Various deposition techniques can be used to provide good ceramic-metal contact:

(a) Sputter deposition or silver evaporation

(b) Screening with silver-based ink followed by heat treatment at 850°C to ensure proper adhesion of the metal layer to the ceramic
(c) Sintering of platinum wires with various metal oxides

Changes in resistivity have been observed for the different deposition techniques. However, these changes are primarily associated with the heat treatment method used rather than the nature of the electrode.

3.2 Thermistor Types

Figure 8[5] illustrates some of the different types of NTC thermistors. The shape and performance of the thermistor can vary with the application.

(a) Top and bottom terminations

(b) Wraparound terminations

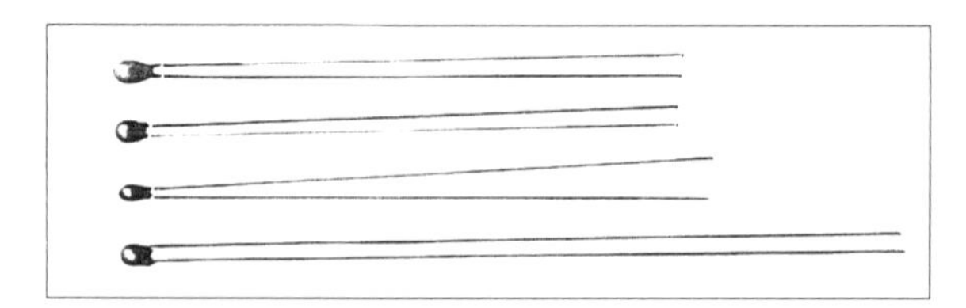

(c) Models M, C, F, T, B NTC thermistors—conformal coated

Fig. 8 Styles of NTC thermistors and assemblies (Dale).
(continued on next page.)

[5]Dale Electronics, a division of Vishay, *NTC Thermistor Catalog,* vol. 7, Rev 919.

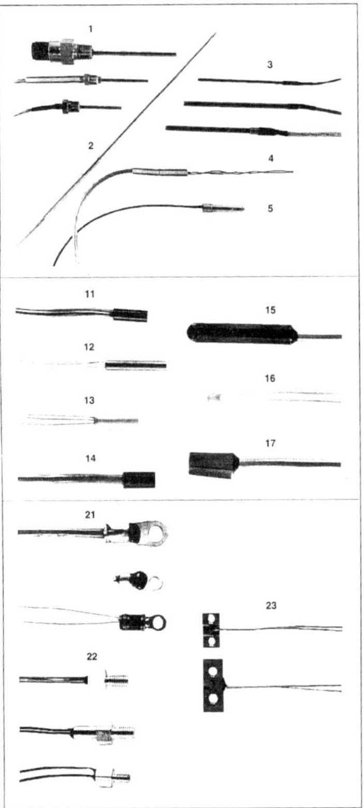

NOTE: A variety of wire styles is available from stock. Depending on the probe and application, the wire will vary. Consult the factory for this and other information.

HOW TO ORDER

1. Choose Style: Immersion probe.
2. Select Style Type: "1" = thread mount.
3. Select Thermistor: 1M1002-C3.
4. Indicate style, type and thermistor to factory for part numbers.

Standard and custom assemblies are available in a variety of configurations. The choice of assembly style is dependent on the application. The primary factors which determine the optimum configuration of a thermistor assembly are the operating environment, mounting, time response and minimum dissipation constant.

The two factors which do vary considerably with assembly design are time constant and dissipation constant. The time constant will typically be of greater duration in encapsulated thermistors. This is, of course, due to additional mass surrounding the thermistor element; therefore, extending the thermal transfer time. Dissipation constant will also be greater in assemblies. The additional housing mass serves well as a heat sink. Greater power is therefore required to induce self-heating.

Both time constant and dissipation constant will vary with the selected thermistor and housing. Heat transfer properties of the housing, thermistor location, mass and wire type determine these constants. It is recommended to evaluate or consult the factory in applications where T.C. and D.C. are critical.

ASSEMBLY STYLES

• Immersion Probes

1. **Thread Mounted:** Features a stainless steel tube with a hermetically sealed threaded hex fitting. Available in 1/8", 3/16" and 1/4" outer diameter tubes with 1/8" or 1/4" NPT hex fittings. Ideal for pressured, closed systems.
2. **Penetration Probe:** 7" long x 1/8" outer diameter stainless steel tube. Pointed tip easily penetrates soft and semi-frozen materials. Also ideal for soil and ground measurements.
3. **Tubular Stainless Steel:** 1/8", 3/16" and 1/4" outer diameter tubes in variable lengths. Good for universal immersion applications. Adaptable to use with compression thread mounts to give designer variable immersion depths.
4. **Tubular Pyrex®:** 6" long x 3/16" outer diameter glass tube. Excellent for lab use where a chemically inert probe is required.
5. **Flexible Immersion:** 1.5" long x .280" outer diameter aluminum housing. Flexible 105°C PVC tubing to protect wire and thermistor element from the liquid medium. Ideal assembly for intermittent immersion in hard to reach areas.

• General Purpose Sensors

11. **Delrin® Housing:** 1/2" long x .170" outer diameter thermoplastic. Excellent for environmental controls and applications below 100°C.
12. **Stainless Steel Rod:** .875" x .156" outer diameter. Good for environmental chamber measurement where corrosive gasses exist.
13. **Gold Anodized Flanged Sensor:** .630" long x .275" outer diameter with a .298" flange diameter. An easy to mount press fitting.
14. **ABS Housing:** .476" long x .230" outer diameter. Black thermoplastic and configuration suits it well in air flow temperature measurements.
15. **Polyester Housing:** 1.50" long x .280" outer diameter. Heavy thermoplastic case makes it ideal for applications requiring delayed time responses. Ideal for process controls in refrigeration and heating.
16. **Epoxy Tip Probe:** Durable epoxy encapsulation, small size, fast time response, versatility and low cost make this sensor universally accepted. Assembly size will vary with wire and thermistor choice.
17. **Pipe Sensor:** .625" long with a cut-out radius of .250" suits this nylon thermoplastic as an ideal pipe sensor. Extensive use is seen in environmental and water heating/cooling systems.

• Surface Sensors

21. **Ring Tongue Lugs:** Surface mount with a #4, #6, #8 or #10 screw. Excellent for measurement and control of surfaces where fast time responses are necessary.
22. **Heat Sink Sensors:** Available in brass, aluminum or stainless steel with various threads. A durable and practical surface sensor, especially where shock and vibration are present.
23. **Rectangular Block Sensors:** Aluminum blocks for measurement and control of large surface areas. Easily mountable with #6 screws.

Refer to "How to Select an NTC Thermistor" for general design aids in choice of thermistor value, tolerance and R-T curve.

(d) Assemblies

Fig. 8 (continued)

Thermistor shapes can vary from a few dozen millimeters in the largest devices to a fraction of a millimeter for the most common designs. They are sold as disks, bare or protected by a layer of varnish with soldered connections for mounting on printed circuit boards. Other thermistor styles are also available, including threaded sensors for mounting on a metal chassis, or as strips or chips for surface mounting. Table 2 provides the dimensions and characteristics of some typical thermistors.

TABLE 2 Dimensions and Characteristics of Typical Thermistors

Model	Diameter ϕ (cm)	Thickness (cm)	Volume (cm^3)	Dissipation (mW/°K)	Response τ (sec)
Bead	0.050	0.050	$1.0 \cdot 10^{-5}$	0.05–0.03	$4 - 5 \cdot 10^{-3}$
	0.150	0.050	$9.0 \cdot 10^{-4}$	0.20–0.30	2.00–15
Chip	0.125	0.025	$3.0 \cdot 10^{-4}$	1	5.00
Smt Chip	0.625	0.060	$4.0 \cdot 10^{-3}$	2.3	4.00
Disk	0.250	0.125	$6.2 \cdot 10^{-3}$	3.5	9–12
	0.500	0.125	$2.4 \cdot 10^{-2}$	7.0	26
	1.250	0.250	0.15	15	65
	2.500	0.125	0.62	25	150
Strip	1.000	0.300	0.23	2.8	35

3.3 Typical Applications

NTC thermistors have a number of advantages that have made them popular for a wide variety of applications in the automobile, manufacturing, and aerospace industries.

- Wide range of resistance and TCRs
- Wide range of operating temperatures (−50°C to 800°C)
- Very high temperature sensitivity (variation of one order of magnitude for a temperature change of 50°C)
- Mechanical strength
- Low cost
- Small size
- Large number of available styles

3.4 Current-Voltage Characteristics

The fundamental characteristic of an NTC device is shown in Fig. 9.

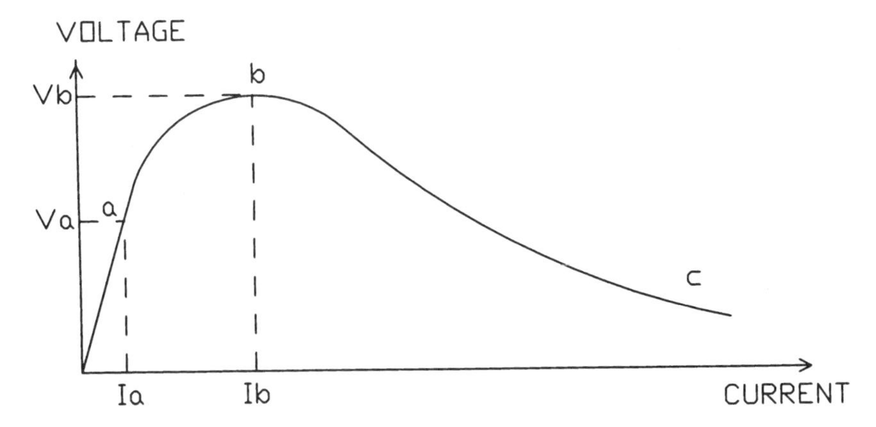

Fig. 9 Voltage vs. current curve of NTC devices.

We will briefly describe several applications that make use of the o-a and b-c regions of the voltage-current curve. Two regions of the curve are typically used in determining suitability to application:

1. **Ohmic region (o-a).** Within this region, the relation between current and voltage is linear. It corresponds to the operating region at low voltages (for example, a temperature probe thermistor).
2. **Nonlinear region (b-c).** This region is characteristic of NTC devices. The voltage decreases when the current increases. This region depends on the heat dissipation factor of the component and its environment (ambient temperature, fluid flow rate, etc.)

3.5 Temperature Sensors (low heat dissipation)

The high TCRs of NTC thermistors (an order of magnitude greater than the most sensitive metals such as platinum) make them highly suitable for temperature measurement applications. The classic example of such an application is the Wheatstone bridge. Under these conditions, the value of the current across the thermistor is found in the o-a region. Thermal contact with the environment enables the device to measure temperature variations on the order of 1/100°C between −50°C and +250°C.

The use of an NTC device within a larger temperature range results in a nonlinear relation between resistance and temperature (end of the o-a region). In this case various mounting technologies, based on serial or parallel resistor arrangements, or logarithmic networks, are used to linearize the thermistor output signal.

3.6 Applications Based on the Nonlinear b-c Region

Many applications are based on modifying the curve shown in Fig. 9 as a function of power dissipation conditions. These conditions will depend on fluid flow variations, liquid immersion, or changes in the nature of the fluids involved. NTC devices are often used to detect abnormal temperature increases (overheating of a motor, ventilation problems, fire, etc.).

For example, Fig. 10 represents the schematic for an alarm circuit that can be used to detect changes in the flow or nature of a fluid. Normal flow is represented by operating point A on the thermistor voltage-current curve. If there is a drastic change in the flow, the operating point moves to B. Detection of the change in current between points A and B can be used to trigger a relay or alarm, or simply to detect a change in the given flow conditions.

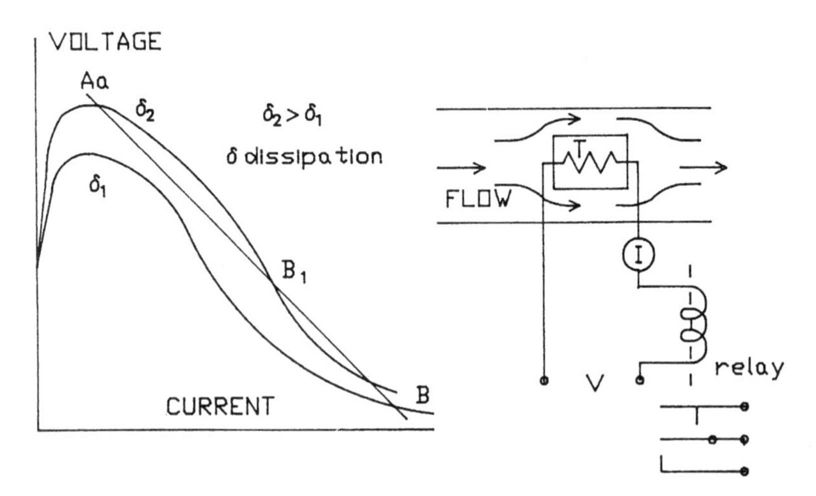

Fig. 10 Schematic of an alarm circuit in a flow of fluid.

3.7 Applications Based on Current-Time Relations

Some applications make use of the time needed for the thermistor to reach its equilibrium point in region b-c. This time depends principally on the thermal mass, the initially supplied power, and its dissipation.

In this section we will describe the use of a thermistor to limit the value of transient currents due to condenser loads in switching power supplies (Fig. 11).

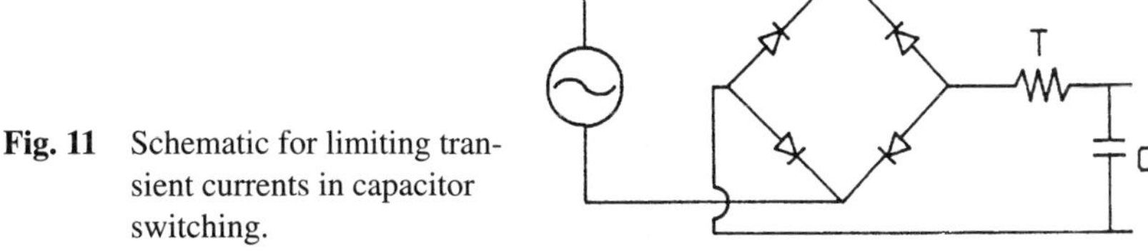

Fig. 11 Schematic for limiting transient currents in capacitor switching.

At $t = 0$ the input voltage is applied, resulting in a charging current in the filter condenser. Given its thermal inertia, the resistor will only heat up after a certain period of time has elapsed (generally a few seconds), thus limiting the value of the charging current. Once equilibrium has been reached (region b-c), the resistance drops significantly and no longer interferes with steady-state operation of the downstream electronic circuit.

4. POSITIVE TEMPERATURE COEFFICIENT THERMISTORS

PTC thermistors are made from semiconductor ceramics based on barium titanate ($BaTiO_3$), which crystallizes into a perovskite structure. Barium titanate is a ferroelectric crystal belonging to the group of crystals with oxygen octahedra. Figure 12 is an illustration of the unit cell.

In the manufacture of PTC thermistors, preparation is similar to that used for conventional ceramics. Titanium oxide, barium carbonate, and other dopants specific to PTC thermistors are mixed and ground in distilled water. After drying, the mixture is fired at a temperature between 1000°C and 1250°C to obtain the perovskite structure. The product is then ground and dried. An organic binder is mixed with the powder to obtain a granulate suitable for pressing. The pieces are then sintered in an oxygen atmosphere at a temperature of approximately 1350°C.

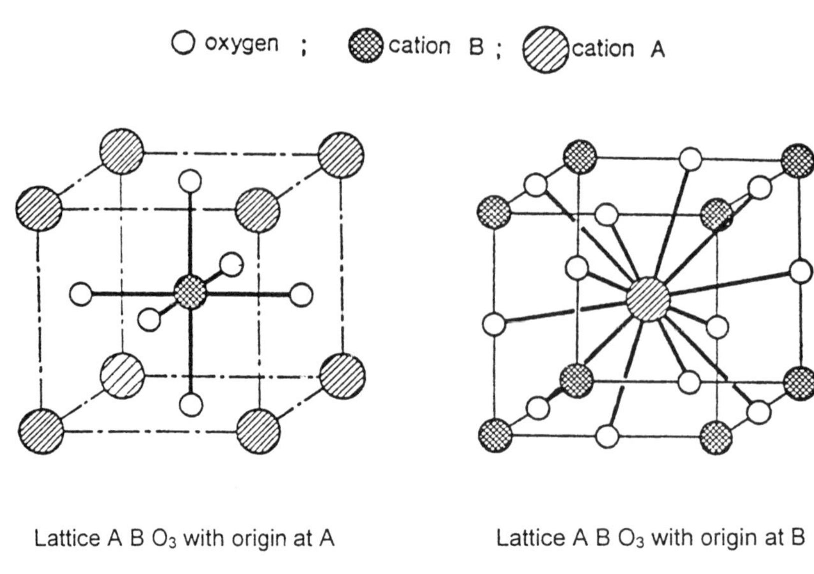

Fig. 12 Unit cell of a barium titanate crystal.

After sintering, electrodes are attached using the techniques described for NTC thermistors. The most commonly used metals are aluminum, nickel, or Al-Ga and Ag-In-Ga alloys. The most common technology is Ag-In-Ga ink screening. The ink contains an oxide that diffuses throughout the ceramic when subject to heat treatment at 600°–900°C. The thickness of the deposit is approximately 10–15µm. *Schooping,* a technique that consists of spraying a molten metal (aluminum, copper, etc.) on the surface of the ceramic, can also be used. The thickness of the deposit is approximately 100µm.

4.1 Chemical Composition and Electrical Properties

The great diversity of PTC thermistor applications requires the development of a number of $BaTiO_3$ dopants. These are primarily characterized by their resistivity at ambient temperature, the Curie temperature at which their resistivity begins to rise sharply, their nominal voltage, and their heat dissipation coefficient.

The resistivity is controlled by the concentrations of donors and acceptors as well as by the cooling profile during sintering, which can be used to control the concentration of barium vacancies at the periphery of each grain. Typical values for resistivity are approximately 20–200Ω · cm. Barium titanate semiconductors are characterized by "*n*-type" properties.

Resistivities of a few Ω.cm can be obtained using a trivalent donor ion (La^{3+}, Ce^{3+}, Y^{3+}), capable of replacing the Ba^{2+} ion. The Ti^{4+} ion can also be replaced by a pentavalent ion (Sb^{5+}, Nb^{5+}) that is also a donor and whose ionic radius is close to that of titanium. Dopant concentrations are less than a few atomic percent.[6]

Although dopants can be used to control the material's resistivity at ambient temperature, they have little influence on the significant variation in resistivity observed in the Curie zone, T_c, which is the actual PTC effect.

This effect, which can result in changes in resistivity greater than 7–8 orders of magnitude within a 50°C temperature range, is associated with the presence of acceptors at the grain boundaries. These regions being relatively thin, electrons are partially or completely compensated by charges that lead to the appearance of potential barriers. This phenomenon was described in section 2. In the case of $BaTiO_3$, the height of these barriers is inversely proportional to the dielectric constant of the material and thus varies considerably with the temperature in the Curie region. The grains, however, remain conductors. Acceptors are often elements that are intentionally introduced near the grain boundaries (Mn, Fe, Cr) or correspond to the presence of barium vacancies at the surface of the grains. Because of polarization, the height of the barriers also depends on the applied voltage. The PTC effect decreases as the applied voltage increases. Two parameters must be taken into consideration to limit this undesirable effect:

1. Increase the thickness of the material to reduce the electric field. Both the geometry and the dimensions of the component can be altered to satisfy this condition.
2. Moderately decrease the grain size to increase the number of barriers in the direction of the electric field. In this case the conditions under which the component is cooled after sintering become extremely critical, and these parameters are among the most closely guarded secrets of the manufacturer of PTC thermistors. Figure 13 is a curve representing the change in resistivity of a PTC resistor with temperature and applied voltage.

The Curie temperature can be shifted through the use of solid solutions of $BaTiO_3$ and $SrTiO_3$ (Curie temperature less than 120°C) or $BaTiO_3$ and $PbTiO_3$ (Curie temperature greater than 120°C). Depending on the materials used, TCR can cover the -50°C to 300°C range. The resistivities obtained are shown in Fig. 14.

[6]Personal communication from A. Lagrange of L.C.C. Thomson, Dijon.

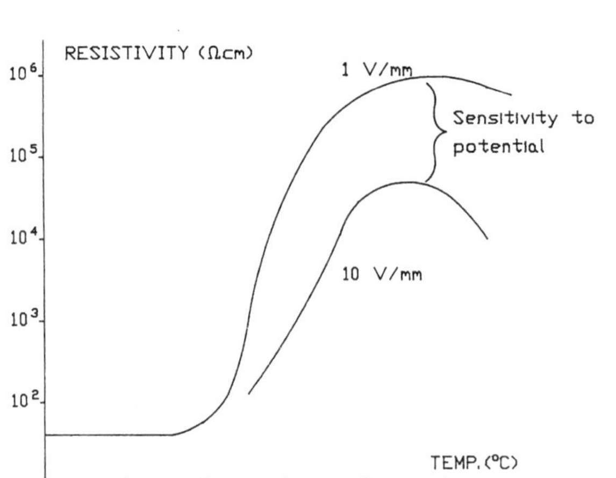

Fig. 13 Resistivity vs. temperature of PTC thermistor for two applied voltages.

Table 3 lists the principal differences among the materials used for NTC and PTC thermistors.

Earlier, we indicated that the characteristics and reliability of PTC devices depend, to a much greater extent than NTC devices, on the chemical treatments and dopants used in their development. Typical PTC chemistry, based on changes in resistivity and Curie points, is summarized in Table 4.

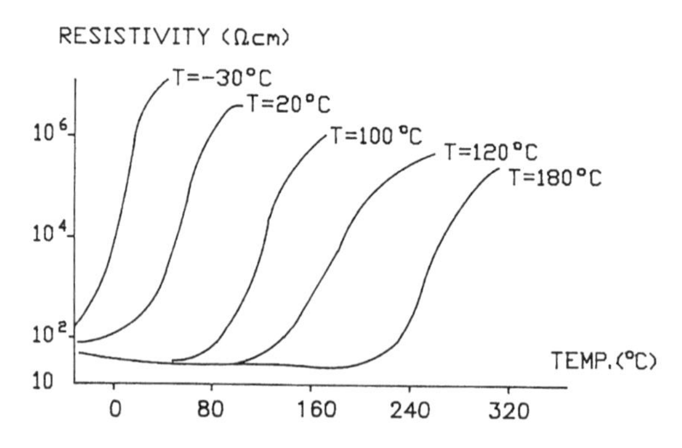

Fig. 14 Resistivity vs. temperature of PTC thermistor for difficult Curie temperatures.

TABLE 3 Materials for NTC and PTC Thermistors

Component	Conductor	General Formula	Material	Crystal Phase	Electrode
NTC	p	$(M_1)^2 + (M_2)^3 + O_4$	$Ni_{0 \cdot 6}MN_{2 \cdot 4}O_4$	Spinel	Ag, Pt, Au
PTC	n	$(M_1)^2 + (M_2)^4 + O_3$	$Ba_{0.995}Y_{0.005}Ti_{1.02}0_3$	Perovskite	Al, Ni, In

TABLE 4 Typical Chemistry for PTC Thermistors

Additive	Valence Change	Purpose of Additive	Substitution Level	Influence on PTC Effect
Na, K, Al, P, Mg	contaminants	impurities	$2 \cdot 10^{-4}$	degradation of characteristics
Si, Ti, Ge	sintering	liquid phase	$5 \cdot 10^{-6}$ to $2 \cdot 10^{-2}$	T° sintering ↓ reliability ↑
Mn, Fe, V, Cu	barrier	grain boundaries	0–0.002	voltage sensitivity ↑ (dR/dT) ↑
Y, La, Sb, Nb	change in valence	change in ρ	0–0.008	control ρ and grain size
Sr, Pb, Ca, Sn	isovalent	solid Ba or Ti solution	0.0–0.2	Curie temperature ↑

5. PTC THERMISTOR TYPES

Given the wide range of uses, PTC thermistors are sold in a number of forms. Common types of PTC (and also NTC) thermistors are surface mount, epoxy beads, wire bond chips, immersion sensors, surface sensors, and others. Some typical designs are described below.

(a) Metallized leadless ceramic. This is the simplest and most widespread design. It is sold in the form of a disk, metallized on both sides, and is designed to be mounted directly between two clips.
(b) Soldered-lead disks. These are also very common in current applications. Two leads are soldered directly to each of the two metallized surfaces used as electrodes. The disk is protected by an epoxy or resin coating.
(c) Ceramic cylinder. Ceramic cylinders are used primarily as heating elements. They are insulated with mineral oxides and mounted in metal tubes to ensure the effective transfer of heat to the exterior.

6. APPLICATIONS

PTC thermistor applications are based on the relation $R = f(\text{T°C})$ and the static voltage-current curve shown in Fig. 15. Here, the PTC thermistor

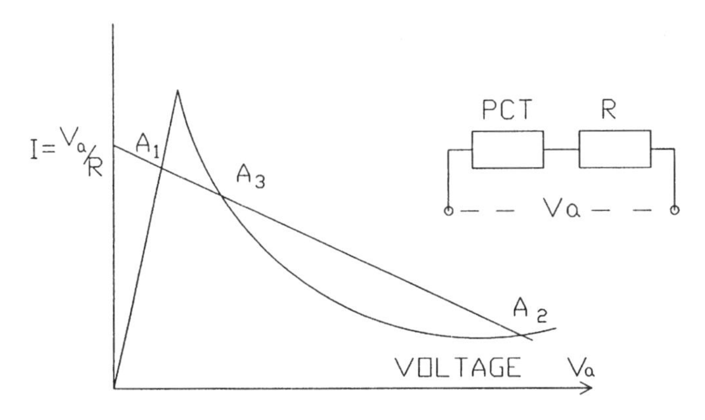

Fig. 15 V-I curve of a PTC thermistor in series with a fixed resistor.

application is mounted in series with a constant load R1. In general, the slope of the load cuts the voltage-current curve at three points, A1, A2, and A3. Only A1 and A2 are stable. The first corresponds to a high current (low R for the thermistor) and the second to a weak current in which the thermistor's temperature is above its Curie point.

This feature can be used to design a variety of protection systems in which thermistors protect the circuit in the event of an abnormal rise in supply voltage. For example, PTC thermistors can be used for alarms and to protect motor windings from overheating.

The thermistors are arranged in the motor so that they are thermally coupled to the windings. In the event of an abnormal temperature increase, the thermistor resistance increases beyond a threshold value, which can be used, together with the use of relays triggered by the PTC, to cut the supply voltage (Fig.16).

PTC devices are also used as heating elements for liquids and gases. In this case the current through the thermistor is high enough to raise the temperature and allow the set-point to be located in the nonlinear portion of the V-I curve. Any change in the type or velocity of the fluid will result in a change in the heat transfer between the thermistor and the fluid. The change in resistance can regulate the quantity of heat transmitted to the fluid, independently of other external conditions.

PTC thermistors are also commonly employed in demagnetization systems in color televisions and videotape recorders. Here, a magnetic field is gradually lowered by means of a coil mounted in series with the PTC (Fig. 17). When voltage is applied, the PTC device is cold and

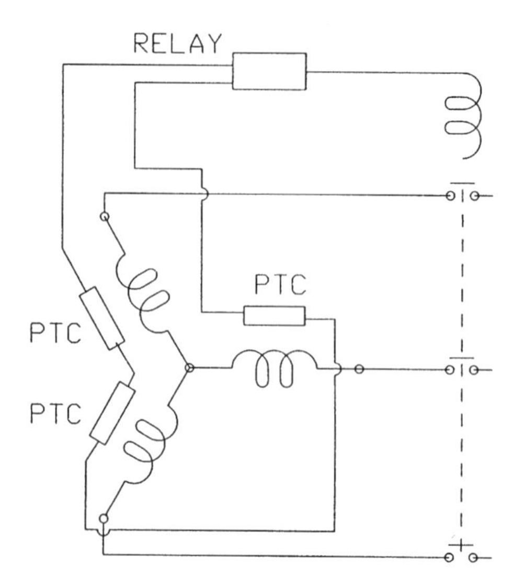

Fig. 16 Schematic of a PTC thermistor protecting the windings of an electrical motor.

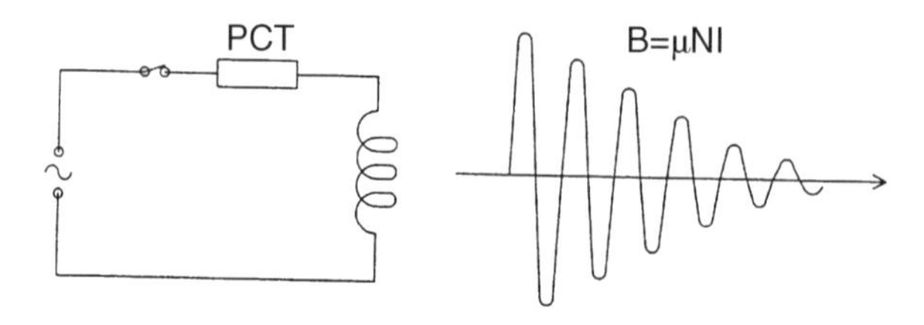

Fig. 17 Demagnetizing circuit.

allows a high-amplitude alternating current to pass through, resulting in a strong magnetic field. Over time, the PTC thermistor heats up and reduces the current and the magnetic field along with it.

The above applications are characteristic of the enormous number of circuits in which PTC thermistors can be used. Although far from complete, Table 5 summarizes some typical applications of PTC thermistors.

TABLE 5 Typical Applications of PTC Thermistors

T°C Measurement	Automotive	Electronics	Other
regulation	fuel control	V/I control	demagnetization
air conditioning	air circulation	protection	heating
fluid flow	carburetor and ignition	engine circuitry etc.	household appliances, hair dryers
	defrosting		

7. VARISTORS

The conductive ceramics used in varistors have properties similar to the $BaTiO_3$ perovskites used for PTC components. Localized grain boundary barriers in zinc oxide crystals here play an essential role.

Zinc oxide is the most commonly used ceramic in the manufacture of varistors. Its crystal structure is identical to that of many other oxides, such as MnO, CaO, etc. It can be considered as resulting from the interpenetration of two face centered cubic (FCC) lattices. In addition to its piezoelectric properties, local deviations from stoichiometry (an excess of Zn^{++} ions) result in the creation of an *n*-type semiconductor with a band gap of 3.1 eV. However, it is the addition of bismuth (Bi_2O_3), cobalt (CoO), chromium (Cr_2O_3), and other oxides—some varistors can contain as many as eight different oxides—that is responsible for the nonlinear variation of current with applied voltage, caused primarily by the effect of these oxides at the grain boundaries.

7.1 Varistor Effects

The voltage-current curve shown in Fig. 18 contains several distinct regions, each of which is characterized by a specific conduction behavior.

(a) *Region I.* This corresponds to a linear relation between current and voltage, with current values being on the order of a few nA/cm^2.

(b) *Region II (pre-avalanche region).* This region corresponds to the normal operating state of the varistor. Conduction (a few nA/cm^2) is heat activated and increases with the applied voltage following a $\log(I) \cong V^{1/2}$ law.

(c) *Region III (avalanche region).* This region corresponds to the actual varistor effect. The current, I, varies nonlinearly with voltage

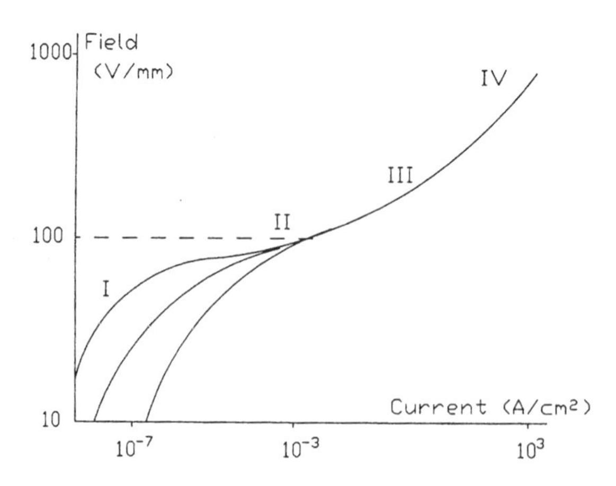

Fig. 18 Electrical field vs. current.

following the relation $I = KV^{\alpha}$, where K is a constant and α the coefficient of nonlinearity. If $\alpha = 1$, the voltage-current relation will predict classic ohmic behavior. When $\alpha \Rightarrow \infty$ the varistor effect becomes perfect. α is generally between 25 and 100 in commercial varistors. These values are provided in the literature supplied by manufacturers. Current values are between a few tens of μA/cm^2 and a few tens of A/cm^2.

The conduction mechanism in this region is due to a tunnel effect, caused by free electrons crossing the potential barrier at the grain boundary. The height and width of these barriers are directly related to the type of dopant used and the applied voltage. This condition occurs very rapidly, on the order of a few nanoseconds.

(d) *Region IV (saturation region).* Above a certain voltage the $I = KV^{\alpha}$ law no longer holds. This corresponds to a region of high current (a few hundred A/cm^2). The current-voltage relationship is quasi-ohmic for very high currents (above 10,000 A/cm^2). Conduction is limited only by the resistivity of the ZnO crystallites rather than by the potential barriers at the grain boundaries.

7.2 Varistor Structure and Fabrication

As indicated above the basic varistor material is ZnO. In its crystal phase the resistivity of the grains is less than $\rho \leq 1\Omega \cdot$ cm at 300°K. The small amounts of other oxides present generally do not exceed 2–3%. A typical composition contains 97% at. ZnO, 1% at. Sb_2O_3, and 0.5% each of at. Bi_2O_5, CoO, MnO and Cr_2O_3.[7]

Varistors are produced using a conventional ceramic process. The different oxides are generally mixed in the aqueous phase. After drying, the powder is heated at 700°C and reground. After the introduction of an organic binder, the powder is pressed in a mold of the desired shape (generally a disk). After heat treatment at 600–700°C to eliminate the binders, the disk is sintered at approximately 1200–1400°C. The disks are then metallized either through deposition of a silver ink or by *schooping*.[8] If appropriate, leads are attached and the entire assembly is protected with an epoxy or phenolic resin coating.

The various dopants are selected not only on the basis of the value of α (in this case the oxides of Bi, Co, Mn, and Ni) but also to satisfy the requirements of grain resistivity, stability, and reliability. This requires the use of different kinds of dopants, whose nature, concentration, and method of application are based on the manufacturer's skill. The microstructure of a typical varistor is illustrated in Fig. 19.

ZnO grains with an average diameter of 10–20 μm are separated from one another by a potential barrier layer whose thickness is a few tens of angstroms. Secondary intergranular phases are present in low concentrations. These are characterized by the following basic structures:

- Spinel $Zn_7Sb_2O_{12}$
- Pyrochlore $Zn_2Bi_3Sb_3O_{14}$
- Bi_2O_3 in α, β, γ, and δ phases

An insulating region, whose thickness is $e \approx 1000$ Å, is also formed around each ZnO grain.

To help analyze varistor phenomena, Fig. 20 represents a theoretical model with a grain size of approximately 20 μm. Grains are separated from one another by an insulating barrier of $e = 1000$ Å. In this structure the critical avalanche threshold, for a current density of 1 mA/cm², is equal to $E_{CR} = 130$ V/mm. The critical potential at each barrier is therefore $V_{CR} = E_{CR} x d = 2.6$ V. In reality, because of dispersion along the dimension of the grains, this value, which is characteristic of the majority of varistors, is

[7]M. Matsuoka, *Japan. J. Appl. Phys.* 10, 736 (1971).

[8]The Schoop process (named after its inventor, who filed the patent for the process in 1910) consists of depositing metals on a surface by spraying them in an electric arc.

Fig. 19 SEM photograph of a varistor showing its cubic structure.

approximately 3.2V per grain. This relation can be used to calculate the thickness needed to produce a varistor with a critical voltage of

$$V_{CRT} = nV_{CR} = \frac{D}{d} \cdot V_{CR} \tag{1.15}$$

where D is the distance between electrodes, and d the average grain size. Typical values for $V_{CRT} = 200$ V are $d = 20\ \mu$m, $D = 1.6$ mm, and $n = 80$.

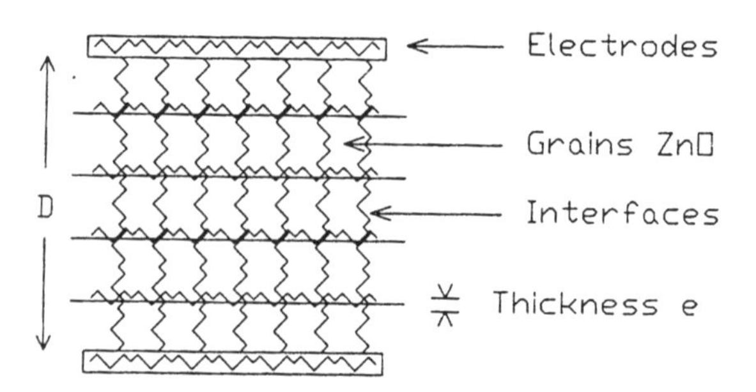

Fig. 20 Physical model of a varistor.

7.3 Equivalent Circuit

Measurement of the dielectric constant of varistors results in values on the order of $\varepsilon \cong 1000$ at 10kH, which is higher than the dielectric constant for ZnO. In fact, the existence of insulating potential barriers at the grain boundaries creates an electrostatic capacitance that results from the basic capacitance of individual grains taken as a whole. Since $d \gg e$, it is obvious that the volume between the component's electrodes is primarily occupied by ZnO conducting grains. For each grain this can be represented by the equivalent circuit shown in Fig. 21.

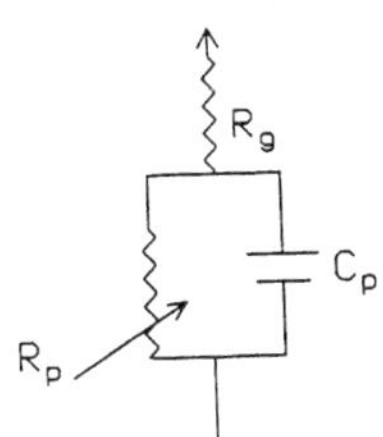

Fig. 21 Equivalent circuit of model shown in Fig. 20.

Here C_p and R_p are, respectively, the capacitance and resistance of the potential barrier, and R_g the resistance of the grain. The resistance R_p is thus parallel to the capacitance C_p. Consequently, the thickness of the insulator between electrodes is not D but De/d. This means that if ε_g is the dielectric constant for ZnO, the total capacitance is then equal to:

$$C = \varepsilon_g \varepsilon_o \frac{S}{De/d} = \varepsilon_g \varepsilon_o \frac{S}{D} \cdot \frac{d}{e} \tag{1.16}$$

Consequently, the measured dielectric constant is greater than that of ZnO by a factor of d/e. Since the dielectric constant for ZnO is on the order of 10, this implies that $d/e \cong 100$. Since $d \cong 10\ \mu\text{m}$, $e \cong 1000\text{Å}$.

The dissipation factor, $t_g\delta$[9], varies with frequency between 0.02 and 0.04, with a peak at 0.10 around 300 kHz. The parallel resistivity of the varistor, ρ_P, is inversely proportional to the frequency ($\rho_p = 1/\varepsilon\varepsilon_o\omega D$) and is essentially equal to 10^9 Ω-cm at 30 Hz and 10^3 Ω-cm at 10^8 Hz. It should also be pointed out that C_P, R_P, and $t_g\,\delta$ vary with temperature.[10]

[9]The physical origin of the potential barrier appears to be due to the creation of holes near the intergrain interface, leading to a space charge region with both capacitive C_p and conductive ρ_P properties. See G. E. Pikes *et al.*, *J. Appl. Phys.* 57, p. 5512 (1985).

[10]L. M. Levinson and H. R. Phillips, *J. Appl. Phys.* 47, 1117 (1976).

7.4 Applications

The static voltage-current curve shown in Fig. 18 can be used to characterize the operation of the varistor in its rest and avalanche regions. The voltage corresponding to a current of 1mA is generally used as a reference point to determine manufacturing tolerances. Beyond this limit characterization is conducted with pulses, using current pulses with a rise time of 4μsec and widths of 10–20 μsec, depending on the densities of the injected currents, which can reach 10k Λ.

The use of transients of increasing value leads to a gradual change in the V-I characteristics, as shown in Fig. 22. This phenomenon is general and valid regardless of the shape of the incident wave. Two regions remain unchanged: the region corresponding to conduction arising from tunneling through barriers (V_{CR} is invariant) and the region corresponding to high currents, where the conduction mechanism is controlled by the resistance of the ZnO grains. The pre-avalanche zone, however, is modified in the direction of a gradual increase of the rest current $I_1 \Rightarrow I_2$.

The physical mechanisms involved vary depending on the duration of the pulse, which can range from a few microseconds to a few milliseconds. In the first case the temperature increase is localized at the grain boundaries and can lead to the creation of cracks from local heat effects. In the second case the temperature of the material rises. If the electrical power is significantly higher than the allowed value, we can observe a breakdown of the material. This results from local melting caused by the presence of Bi_2O_3 phases with a melting point of nearly 800°C.

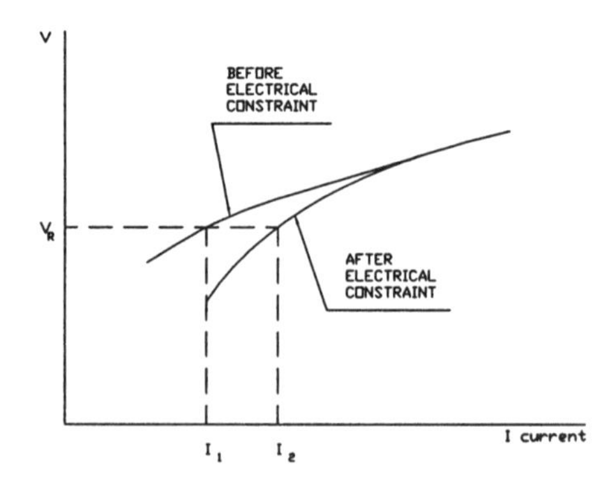

Fig. 22 Effect of electrical constraint on V-1 curve (V_R-rest voltage).

7.5 Varistor Types

Varistors are generally manufactured in the form of disks whose thickness and diameter vary with the application (from 5 to 80 mm for the diameter and from 1 to 50 mm for the thickness). The sides are protected by an organic or vitreous coating that prevents interelectrode flashover. Typical values for the different parameters are shown in Table 6.

TABLE 6 Parameters of Varistors

Rest voltage	10–10,000 V
Rest current	1–100 μA
Response time	$<$ 50 nsec
Capacitance	50–12,000 pF
Allowable peak current	6000 A/cm^2
Allowable transient energy	400 J/cm^2

7.6 Areas of Application

7.6.1 Medium and high voltage

The most typical use for varistors is to protect electrical lines and transformer stations against lightning. The transient currents and energies propagated by conductors are very high and can reach 150 kA and $2.5 \cdot 10^6$J, respectively. During the past few years, spark-gap lightning arrestors on high-voltage lines (400 kV and up) have been replaced with ZnO varistors. These devices consist of a stack of varistor disks in series, the number of disks being proportional to the rated operating voltage. For example, a protection system for a 400 kV network requires a stack containing hundreds of disks nearly 3 meters high.

7.6.2 Low voltage

For low-voltage (50–1000V) applications, varistors are used as circuit protectors for telecommunications, computer systems, and mass-market and industrial devices, in which they provide very effective protection against pulses propagated by power circuits. A 30-mm diameter varistor can protect against transients up to 30 kA.

Another significant source of voltage spikes results from switching inductive and capacitive loads on low-voltage networks. These can propagate over long distances and lead to the destruction of components, the deterioration of relay contacts, and improper operation of logic circuits (CMOS and others), which are especially sensitive to transients.

Compared to atmospheric discharges, noise from industrial sources results in transients of short duration and low magnitude, but their repeated action can produce significant cumulative effects. For example, opening and closing the relays on an inductive load can lead to voltage peaks that are 10 to 50 times greater than the applied voltage. Similar results have been found with switching on the primary coils of transformers. Varistors also provide effective protection against electrostatic discharge, the primary cause of damage to MOS junctions.

For very low voltage applications (less than 50 V), such as those in the automotive industry, different sources of disturbances (starter, relays, windshield wipers, alternator) lead to voltage peaks that can reach hundreds of volts and interfere with the proper operation of electronic circuits.

8. GAS SENSORS

Gas sensors represent another family of nonlinear resistors; a brief description is provided below. Their operation is based on Redox reactions in ionic conductive ceramics[11]. The conductive charge density is dependent on the concentration or partial pressure p of gas absorbed (oxygen, for example) by the ceramic, modifying the conductivity according to the following equation:

$$\sigma = A(T^o)p^n_{o_2} \quad \text{in which} \quad A(T^o) = A_o \exp(-E/kT)$$

controls the ionization and crystal defects, and n varies between $\pm 1/4$ and $\pm 1/6$. A positive exponent corresponds to a p-type conductivity (excess of oxygen), whereas a negative exponent results in n-type conductivity (lack of oxygen). The two ceramic oxides that have been most frequently studied in this field are compounds of the type $Co_{1-x}Mg_xO$ ($x < 0.85$), a p semiconductor, and compounds of TiO_2, which is an n semiconductor. Table 7 lists some of the materials used in gas sensor manufacturing.

[11]There is another type of gas sensor that is based on the existence of electrochemical potentials that follow the Nernst law in zirconium ceramics.

Table 7 Materials Used in Gas Sensors

Gas Sensor	Conduction	General Formula	Material	Crystal Phase	Electrode
Oxygen	Ionic	$(M_x)^2 + (M_{1-x})^4 + O_4$	$Y_{0\cdot1}Zr_{0\cdot9}O_{1\cdot95}$	Fluorite	Porous Pt
Humidity	p	$(M)^2 + (M_2)^3 + O_4$	Ti doped $MgCr_2O_4$	Spinel porous	Porous RuO_2
Hydrocarbon gas	n	$(M)O_2$	Doped SnO_2	Rutile	Porous Pt

One of the major defects of these sensors, which are based on nonlinear resistance, is the significant change in resistivity with temperature. For TiO_2 the TCR at 400°C is on the order of $4 \cdot 10^{-2}$/°C. Their use in automotive applications (to detect the presence of oxygen) requires that they be preheated above normal operating temperatures to overcome these changes.

9. TIME CONSTANTS FOR TEMPERATURE MEASUREMENT

The time constant to reach thermal equilibrium with their environment is an important characteristic for thermistor use. The order of magnitude for the constant is estimated as follows. Assume a thermistor of typical dimension L (average dimension), specific heat C_p, and heat transfer coefficient k. This thermistor is placed in a fluid so that the transfer coefficient from convection and radiation is h.

We can demonstrate that, depending on the value of the Nusselt number[12] $\mathfrak{N} = hL/k$, either the heat resistance of the component or the boundary layer between its outside surface and the fluid will control the heat transfer process. In the case of a thermistor where $L \cong 3$–10, $k = 1$–2 W/m°C, placed in air or gas, h is approximately 6–570 W/m²°C, depending on whether the convection is natural or forced.

Under these conditions $\mathfrak{N} = 10^{-1} - 10^{-3}$ and the boundary layer controls the heat transfer process. We can then write the following equation:

$$d_q \text{ (heat transfer)} = Sh(T_1 - T)dt = V\rho_d C_p dT$$

in which S is the active surface of the component, ρ_d its density, and V its volume. Integrating the equation yields the following (T_1 is the temperature of the fluid and T_2 the initial temperature of the thermistor):

$$t = \frac{V}{S} \cdot \frac{\rho_d C_p}{h} \cdot Ln\left[\frac{T_1 - T_2}{T_1 - T}\right] \qquad (1.17)$$

The time constant τ_{TC} is obtained when the thermistor temperature is equal to $T_1 (ln = 1)$. For the majority of thermistors, the order of magnitude

[12]The Nusselt number is a dimensionless number used in fluid mechanics. It is the ratio of the amount of heat exchanged by convection to the amount of heat exchanged by conduction. h is the conduction transfer coefficient and L the ratio of a reference length of the solid being studied and the thickness of the thermal boundary layer. This number is also known as the Biot number. See S. Whitaker, *Fundamental Principles of Heat Transfer,* chapter 4, Pergamon Press.

of the constant varies between a few milliseconds and 2 seconds. When the thermistor dissipates heat, the equations given in Chap. 9 for power resistors apply.

10. CONCLUSION

Nonlinear resistors, which were a laboratory curiosity no more than 20 years ago, have become one of the largest growing members of the passive component family, with annual growth exceeding 10 percent a year. The demand for increasingly responsive control systems will continue to fuel this growth. It more than justifies the efforts of those now working in private industry and university laboratories to develop increasingly sophisticated ceramics. These efforts are primarily geared toward obtaining improved sensitivity and reliability in the face of external constraints. This goal has become even more significant since the quality/price ratio for these components is superior to other components with similar features. Moreover, the need for energy savings has given rise to new methods of synthesizing ceramics, foremost among which are sol-gel techniques.

Appendix I

A Precision Film Resistor Exhibiting Bulk Properties

A.1. INTRODUCTION[1]

For many years the engineer requiring a precise (±0.1% or less) resistor has used precision wirewound products. These can be made to tight tolerances, have good stability and are particularly useful in network arrays. Because of their coiled wire construction, they have poor high frequency performance particularly above 50 kc/sec. The reliability of these resistors has been quite variable in the past. They also have poor wattage and resistance range to size relationships. Nevertheless, these resistors have been the standby for most of the close tolerance resistor requirements.

Evaporated metal film resistors have been developed extensively in the last fifteen years [1–7]. They have offered somewhat improved reliability, versatility, and excellent high frequency performance. Their ac-to-dc resistance ratio remains close to unity up to the 100 megacycle/sec region. These resistors have replaced wire types in many applications because of smaller sizes, and in the premium grades (T-9) have reasonably

[1]This chapter is a transcript of a paper coauthored in 1963 by Dr. Felix Zandman, Vishay Instruments, Inc., Malvern, PA and Dr. Sidney J. Stein, EMC Technology, Philadelphia, PA. It describes the foundations of the foil resistor technology.

small temperature coefficients (*TC*) of resistance (±25 ppm/°C). However, the temperature coefficient changes of any two resistors will randomly fall into the total 50 ppm/°C span of such a product. The environmental resistance changes shown by these evaporated film resistors can cumulatively add up to greatly exceed their initial tolerances. Thus, resistance changes can easily reach ± (1% or 2%) regardless of whether the initial resistor tolerance might have been ±0.1% or 1.0%. When changes amounting to ±2% occur, then a total design tolerance of 6% would have to be considered in a circuit employing ±1.0% units or 4.2% for ±0.1% units. This degree of instability has prevented evaporated metal film resistors from being used as a direct replacement for all precision wire types.

This instability is due to the problems inherent in extremely thin metastable film systems. These thin films are so different in characteristics from their bulk metal counterparts as to really behave as rather different materials. This has been attributed to various factors including lattice distortions, discontinuous aggregate formation, occlusion of gas at crystal interfaces, oxidation of the film to form oxide semiconductors, mechanical strains and atomic mobility [8–9]. Some evaporated film resistors, under the influence of moisture and polarizing voltages, also can suffer electrolytic attack leading to catastrophic failures.

A need was felt for a resistor which could combine the small size, and high frequency properties of the evaporated films with the accuracy, precision and stability of good wire types. Such a resistor has been developed at Vishay Instruments, Inc., and will be described here.

A.2. DESIGN CONCEPTS

Meeting the desired objectives of optimizing the characteristics of both metal film and wirewound resistor types requires at least the following:

1. The use of a thick enough alloy film to have bulk properties.
2. Special preparation of alloys so as to produce a low temperature coefficient of resistance.
3. Making these alloys reproducible from batch to batch.
4. Careful control of crystal structure, strain, substrate characteristics, deposition techniques, stabilizing treatments, and all other factors which might lead to subsequent resistance changes.
5. Utilization of the alloys and substrates in a way that limits and controls thermal and mechanical strains either in manufacture or use.

6. Selecting all other parts of the resistor structure to minimize electrical, thermal, chemical, and mechanical stresses. [10]
7. The use of a flat film construction to keep inductance and capacitance low.
8. Processing resistors in large groups and applying uniform treatments to all products up to the final resistance tolerancing to insure good tracking characteristics.
9. Designing a tolerancing method which does not degrade resistor qualities.

A.3. PRINCIPLES USED IN THE RESISTOR CONSTRUCTION AND MANUFACTURE

A metal film of appreciable thickness (30 to 150×10^{-6} in) is deposited on a large area dielectric substrate. The substrate may be rigid or flexible, i.e., both plastic and glass types have been used. In the standard (S-102) construction shown in Fig. 1, glass is used as the substrate.* High thermal conductivity metal substrates have also been used for higher power applications.

The surface of the resistance alloy is coated with photoresist, and a multiple image of the desired resistor pattern is printed on the photoresist.

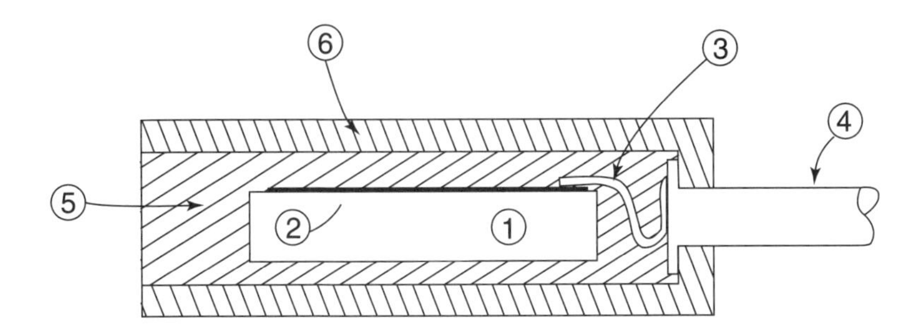

(1) Substrate
(2) Resistance Film
(3) Contact Ribbon
(4) Lead Wire
(5) Encapsulant
(6) Case

Fig. 1 Vishay style S-102 resistor cross-section.

*Presently, ceramic is used as a substrate as it exhibits better heat dissipation characteristics.

After developing the photoresist, the undesired metal areas are etched away. It is very important to use negatives of the best obtainable quality for image printing. High resolution control of uniform resistive path geometry is essential. The dimensional variations along the edges of the etched current path must be kept small as compared to the film thickness, distance between adjacent paths, and width of the path. Because of the relatively low ohms per square value of our bulk property films, very close etching is required to achieve long path lengths. Cleanliness of line edge and spaces is required. After etching, individual resistors are cut from the large substrate. Metal contact ribbons are welded to termination tabs and wire leads thus providing a flexible connection.

Resistors are adjusted to final tolerance by additional metal removal or additional cutting of individual lines in the resistive path. Figure 2 illustrates one type of resistor line pattern showing coarse and fine path adjustment possibilities. This resistor was designed to cover the range of 5 ohms to 1000 ohms in 1/2 ohm steps. Adjustment is performed by extending the etched line to the edge of the resistance element. The letters on this pattern are 0.003″ high.

After tolerancing, the resistors are encapsulated in a molded epoxy shell or coated with a strong moisture-proof resin. At a number of stages during the manufacturing process, stabilizing treatments and quality control tests are performed.

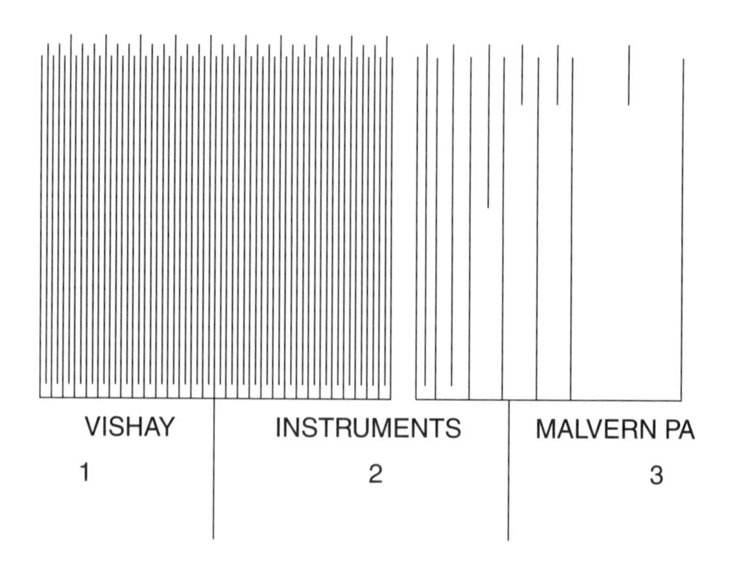

Fig. 2 Resistor adjustment pattern.

A.4. SOME STRAIN CONSIDERATIONS

Fine control of temperature coefficient (TC) is accomplished by mechanical compensation for inherent resistivity change in the alloy. As specific resistivity of the alloy changes with temperature, the dimensions of the relatively massive substrate vary also by thermal expansion. The superimposed thin resistor element is constrained to follow the deformation of the substrate. These two temperature-sensitive effects occur simultaneously, and by balancing the resistance changes inherent in the temperature coefficient of resistivity of the alloy with the resistance changes caused by mechanical strains, the net temperature sensitivity of the integral resistor package could be reduced to zero if both effects were linear, equal in magnitude, and of opposite sign. Thus, resistor alloys must be uniquely matched to substrate materials. Since the resistivity change with temperature of the alloy is not linear, the matching should be made so that the resulting resistance change will vary a minimum amount around a chosen temperature, for example, 24°C. With proper centering of this $\Delta R/R = f(t)$ curve, a TC of zero ±1 ppm/°C, between 0°C to +60°C, is obtainable.

The existence of internal stresses in the resistor leads to the consideration of stability and TC control. In order for the above discussed properties to remain completely reversible and repeatable, stresses in the resistor film and the substrate must be predictable and should not exceed their respective yield points. The materials used in the style S-102 resistor have the following characteristics:

Modulus of elasticity of the metal film and substrate, respectively:

$$E_m = 31 \times 10^6 \text{ psi}$$

$$E_s = 10 \times 10^6 \text{ psi}$$

Yield point of the metal film:

$$\text{Yield point} = 120{,}000 \text{ psi}$$

Poisson's ratio for both materials:

$$\mu = 0.3 \text{ (approx.)}$$

Coefficient of thermal expansion of the metal film and substrate, respectively:

$$\alpha_m = 11.5 \times 10^{-6} \text{ in/in/°C}$$

$$\alpha_s = 8.5 \times 10^{-6} \text{ in/in/°C}$$

Both materials behave like Hookian elastic bodies.

Because the substrate is usually more than 1000 times thicker than the resistor film, and since E_s and E_m are close to the same order of magnitude, thermal stresses in the substrate due to the difference in coefficient of thermal expansion of both materials would be entirely negligible.

For a change in temperature, $\Delta T = 1°C$, the thermally induced strain

$$(\Delta l/l)_T = \varepsilon_T = (\alpha_m - \alpha_s) \cdot \Delta T$$

$\varepsilon_T = (11.5 - 8.5) \times 10^{-6} \times 1 = 3.0 \times 10^{-6}$ inch/inch.

For the entire operating temperature range, $\Delta T = 180°C$ (−55°C to +125°C), the strain will be $180 \times 3.0 \times 10^{-6} = 540 \times 10^{-6}$ inch/inch.

The corresponding thermal stress in the metal film σ_m can be expressed by:

$$\sigma_m = \frac{E_m}{1 - \mu_m} \int_{T_{\text{initial}}}^{T_{\text{final}}} (\alpha_s - \alpha_m)\, dT$$

$$= \frac{E_m}{1 - \mu_m} \times \varepsilon_T = 23{,}900 \text{ psi}$$

or 1/5 of the yield point. Thus, such thermally induced stresses are safely below the yield point and do not cause plastic flow in our bulk metal film.

A.4.1 Resistor Characteristics and Test Data

Use of the precision photoetch techniques and adjustment methods mentioned above provides the basis for extremely tight tolerancing of these resistors. Initial tolerances of 0.01% are produced as standard units, and matching to 0.005% can be achieved.

A.4.2 Temperature Coefficient*

As described previously, the TC may be selected by proper choice of resistance alloy and substrate. Figure 3 shows the temperature-resistance curve for alloy K3. K3 has an average TC of 0 ± 5.0 ppm/°C regardless of resistance value, and tracking of any two resistor values to 0.5 ppm/°C over the span of −55°C to +125°C. The TC for films of another alloy (C3) average 0 ± 7.0 ppm/°C for the −55°C to +125° range with tracking of 0.2 ppm/°C for any two resistors. Resistance range coverage is not as broad for C3 as for

*Presently, much better performance characteristics of foil resistors have been developed. For example: TCR of ±0.4 ppm/°C over the MIL temperature range.

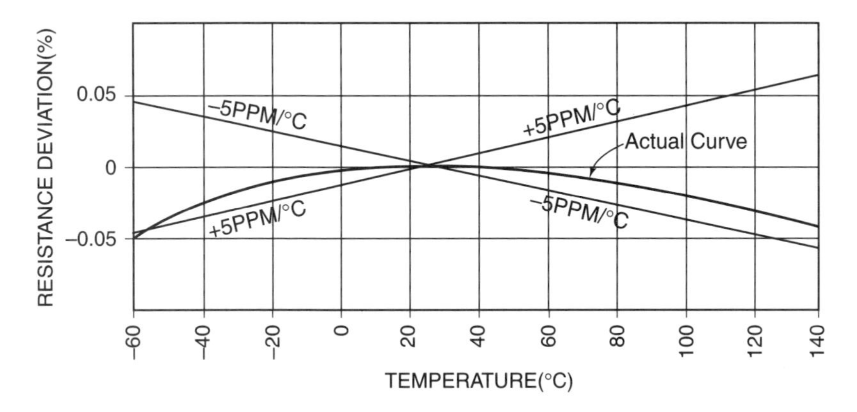

Fig. 3 K3 resistance-temperature curve.

K3 because its ohm per square value is lower. Of particular interest is the chosen location of the maximum in the temperature-resistance curve (TC centering). It is reproducibly set at 24°C so that the lowest TC is obtained in the most useful temperature range for "normal" operation.

TC tracking of ±1 ppm/°C has been observed down to cryogenic temperatures (liquid nitrogen). Close tracking also is observed to temperatures well above 125°C.

A.4.3 MIL-R-10509D Char. E. Test Data

The table in Fig. 4 compares requirements of MIL-R-10509D (Char. E.), MIL-R-93C and typical results of the Vishay-EMC resistor. Performance levels far exceed requirements. Extensive additional testing is in process.

A.4.4 Stability

Six months' shelf stability is shown in Fig. 5. Changes are expected to be in the order of 25 ppm/yr. Figure 6 shows the effect of power at 25°C for ten individual S-102 resistors. Even at a 1-watt (overload) level, resistor stability is still excellent after 100 hours. Eight of the resistors show changes of less than 0.01%. After initial changes of 0.02% and 0.04%, respectively, the remaining two resistors showed no additional change.

An added feature is the combination of long-term stability with shock and vibration resistance. The normal precision resistor, designed for long-term stability, employs loosely wound wire to avoid thermal and mechanical stresses. Such a unit is prone to give difficulty in shock or vibration

	Limits MIL-R-93C (Wirewound)	Tightest Characteristic Limits MIL-R-10509D (Metal Film)	VISHAY-EMC S-102
Temperature Coefficient	0.003%	0.0025%	0.0005%
Moisture Resistance	0.250%	0.5000%	0.0500%
Short Time Overload	0.100%	0.2500%	0.0100%
Temperature Cycling	0.200%	0.2500%	0.0250%
Dielectric Withstanding Voltage	0.050%	0.2500%	0.0080%
Terminal Strength	0.050%	0.2500%	0.0050%
Load Life	0.500%	0.5000%	0.0020%
Low Temperature Storage	0.200%	—	0.0050%
High Temperature Exposure	0.500%	—	0.0200%
Low Temperature Operation	0.250%	0.2500%	0.0050%
Effect of Soldering	—	0.1000%	0.0050%
Insulation Resistance	—	10,000 MΩ	750,000 MΩ

Fig. 4 Precision resistor comparisons.

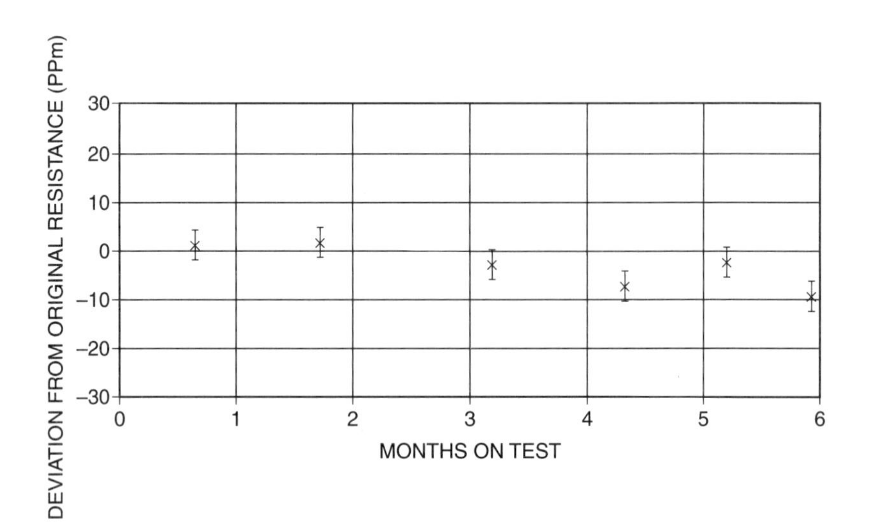

Fig. 5 Shelf stability—room ambient uncontrolled.

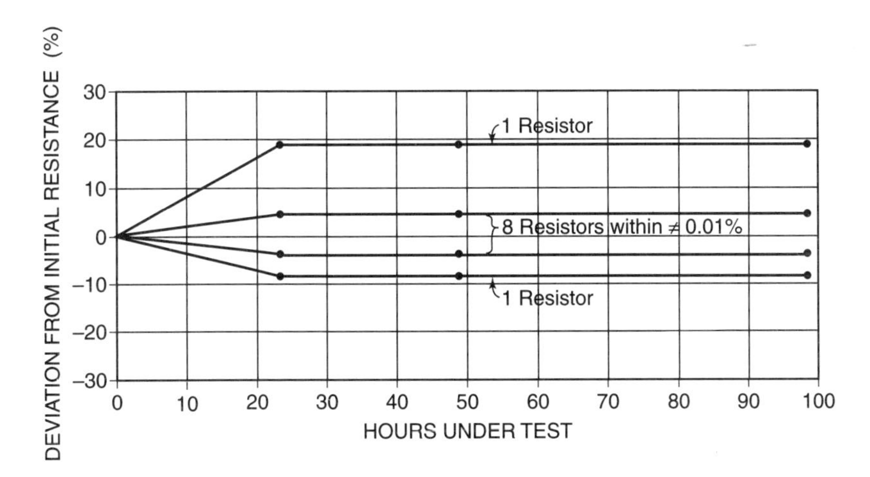

Fig. 6 Life test data—1-watt style S-102.

testing. The Vishay resistor, however, with its monolythic construction, is ideal for such high mechanical stress situations.

Another example of the monolythic structure advantage is the hydrostatic pressure effect: The glass substrate resistor will change resistance reversibly with pressure at a rate of −1 ppm/for each change of 20 psi in pressure. This effect is seen up to at least 50,000 psi. At this pressure the ΔR will be a total of −2500 ppm. At greater than 50,000 psi hydrostatic pressure, permanent resistance shifts are possible.

A.4.5 Power Handling

Figure 7 shows a 1500 hour load life curve at 125°C applying both 0.1 and 0.3 watts. Sample size was 10 units. It may be generally stated that maintenance of film temperatures below 175°C will result in excellent long-term behavior. S-102 film temperatures increase approximately 25°C for each 0.1 watt applied. Special heat sinking, including use of metal substrates, can greatly increase power dissipation capabilities. The table in Fig. 8 lists the sizes, styles, ranges, and power ratings of three standard resistors. Figure 9(a) is a photograph of these resistors.

A.4.6 High Frequency Characteristics

Being a flat film resistor, the basic unit is essentially noninductive. It shows, however, an equivalent parallel capacitance of 0.22 MMF for a 5K

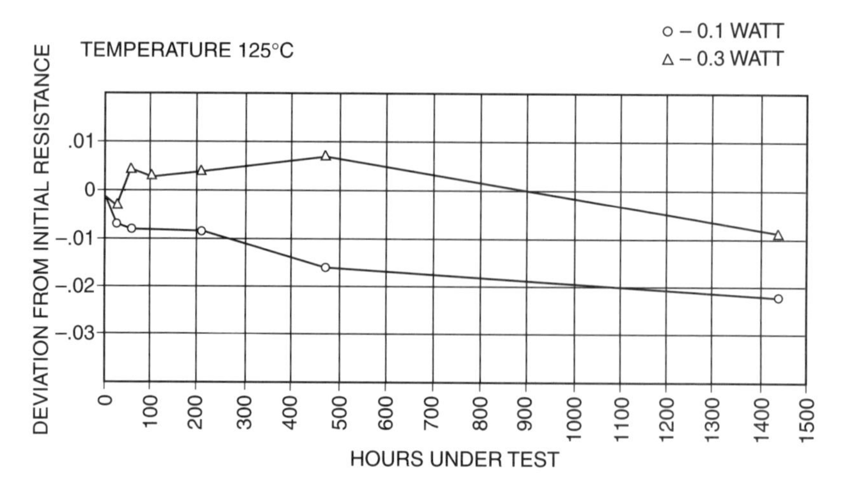

Fig. 7 1500 hour load-life test data, style S-102.

resistor as measured at 100 megacycles per second. The distributed capacitance effects are so small as to permit the resistor rise time to be of the 1 nanosecond order of magnitude.

Special photoetch patterns have been used to improve these properties and to make flat (.002″ thick) unencapsulated stripline resistors for use in the microwave region up to 2000 mc/sec. The same resistors deposited on metal substrates will exhibit a rise time of approximately 50 nanoseconds.

Resistor Style	S-102	D-104	D-105
Dimensions:	.3″ × .3″ × .1″	.56″ × .37″ × .12″	.81″ × .43″ × .12″
Lead Spacing:	0.15″	0.44″	0.69″
Resistance Range:	100 Ohms-30K	200 Ohms-60K	300 Ohms-100K
Tolerances:	All styles available as 0.01% to 1.0%		
Max Voltage (DC)	200 V	300 V	600 V
Power Rating at 125°C			
0.25% and 1.0% Tol.:	.3 W	.4 W	.5 W
0.05% and 0.1% Tol.:	.22 W	.3 W	.37 W
0.01% and 0.02% Tol.:	.15 W	.2 W	.25 W
Derating to 0 power at 175°C			

Fig. 8 Product specifications.

A.4.7 Noise

Noise was measured on a Quan-Tech Model 2136 instrument. In all cases the resistor current noise was below the inherent noise level of the measurement equipment. Using correction curves the following data were calculated:

20 ohm units—less than −33 db

5000 ohm units—less than −43 db

A threefold increase in test current for the 5000 ohm units gave no measurable increase in current noise.

A.4.8 Voltage Coefficient

If temperature of resistor is maintained constant, the voltage coefficient is essentially nonmeasurable.

A.4.9 Reliability*

While reliability information is not yet available, the monolythic design and all welded construction should prove this unit superior to those utilizing pressure contacts, paints, cements and solders.

A.5. LIMITATIONS

The bulk metal film resistor is relatively new, having been developed only during 1962–1963. At the present time,[2] only relatively few styles and restricted ranges are available. Similarly, long-term performance data do not yet exist.

A.6. OTHER APPLICATIONS

This resistor is adaptable to a great number of special applications due to its size and performance characteristics. Principally, any type of network can be assembled from individual resistance subassemblies or chips which

[2]Time of publishing of this paper, 1963.

*Presently, Vishay foil resistors are qualified to the MIL Established Reliability R level (RNC/Y) class.

are then mounted on a common printed circuit board. Absolute tolerances of the individual resistors and resistance ratings can be made to high precision. Because of the inherent TC tracking characteristic, the resistance ratio will be maintained over the operating temperature range. Figure 9 shows a binary ladder network prior to final encapsulation. The actual size of this network is 1.750″ × .750″ × .300″. It contains 22 chips. It is a back-to-back, two-layered construction with 11 chips per layer. Only the inherent stability of the resistance chips makes the manufacture of such a network feasible. These subassemblies remain sufficiently stable so that production of 0.01% tolerance networks becomes practical.

The same inherent stability lends itself to special applications as a bondable resistor. Resistors deposited on flexible substrates 0.001″ thick and approximately 4/10″ × 2/3″ or 1/4″ × 3/16″ can be bonded to any desired structure or heat sink and then adjusted. In essence this resistor could be used as a "one-way" trimmer. Such resistors have been bonded to chassis, potentiometer cases, ceramic spacers, and other parts without essential loss in their characteristics. Figure 10 shows a compensation network mounted on a metal cap. Due to the good heat sink characteristics of such metal "substrates," loads of up to 50 watts/sq inch have been dissipated in some applications.

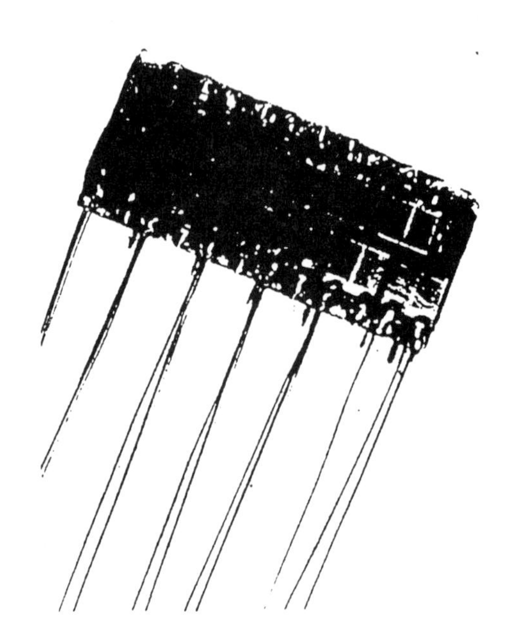

Fig. 9 Unencapsulated 22-resistor ladder network.

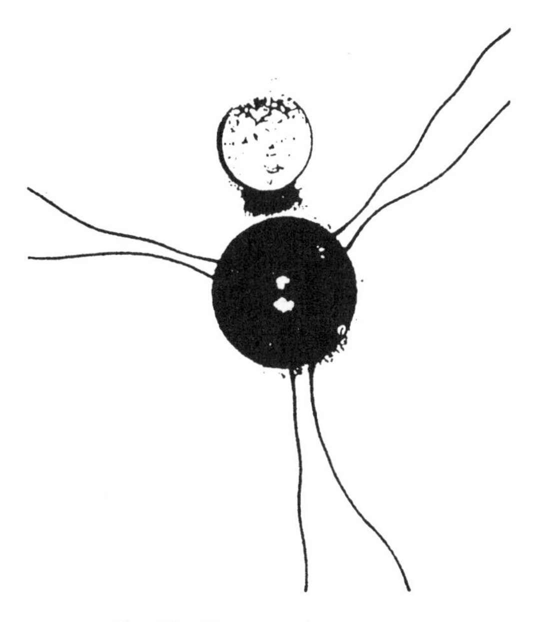

Fig. 10 Compensation network.

Another special application includes half and full bridges. Figure 11 shows a half-bridge, about 3/16″ by 1/16″ size mounted on a steel substrate and consisting of two 350 ohm resistors which track better than 1/2 ppm/°C.

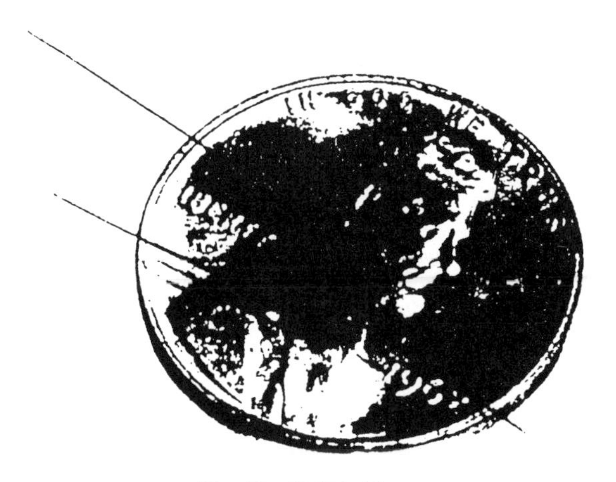

Fig. 11 Half-bridge.

Various metal alloys can be used successfully in the Vishay process, as indicated earlier. Therefore, high temperature coefficient components can also be made. Figure 12 shows a temperature-sensitive Wheatstone bridge circuit. A 5000 ppm/°C resistor in combination with three "zero" TC resistors was used to produce this bridge. This circuit demonstrated a 15,000°F/sec response rate. Note the extremely small size of the entire assembly.

The Vishay-EMC resistance element lends itself to stripline attenuator use. Special configurations give a broadband response and have been used at L-band frequencies. They are made to desired stripline size and can be mounted by means of solderable and weldable lead ribbons.

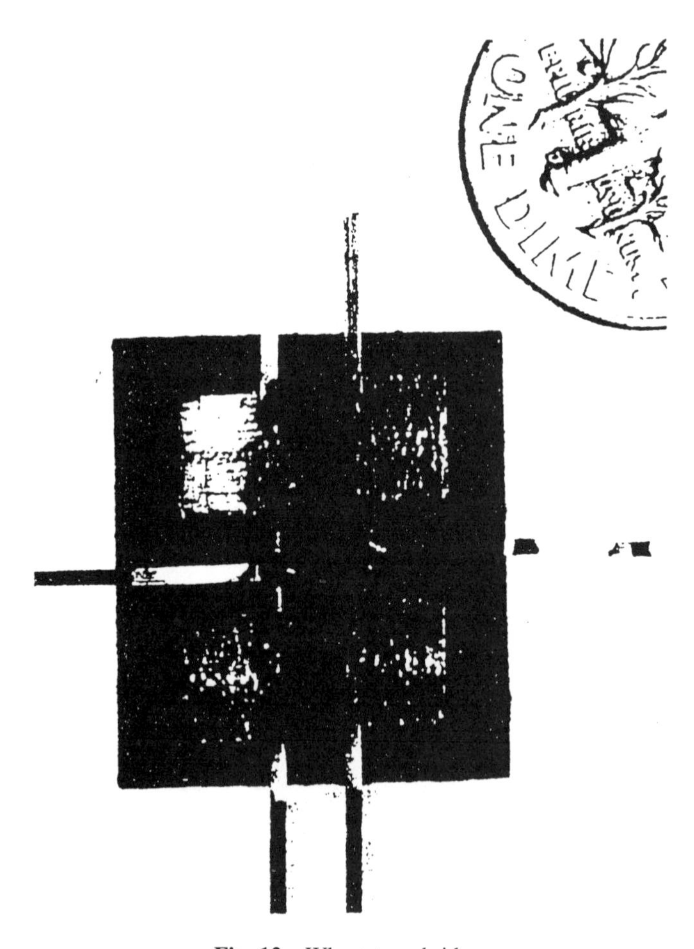

Fig. 12 Wheatstone bridge.

A.7. CONCLUSION

The products developed thus far have demonstrated properties which combine, and in many cases exceed, the high frequency behavior, small size, and ruggedness of evaporated films with the bulk properties, and the TC characteristics, accuracy, and stability of precision wirewound resistors. Resistance ranges are still limited to 100,000 ohms maximum. It is hoped that future developments will make it possible to increase this upper limit, as well as to make other sizes and shapes available.

Additional long-term and larger scale test data are required to give more information on reliability and failure rate performance.

We wish to acknowledge the basic contributions of Dr. D. Post of Vishay Instruments, Inc., to the development of this product and associated techniques.

A.8. REFERENCES

1. C. T. Graham, "Metallic Film Resistors," *Proceedings of Symposium Progress in Quality Electronic Components,* 1952, Washington, D.C.
2. S. J. Stein and J. Riseman, "Evaporated Metal Film Resistors," *Proceedings 1954 Electronic Components Symposium,* Washington, D.C.
3. S. J. Stein, "Multipurpose Evaporated Metal Film Resistors," *IRE, Transactions of the Professional Group on Component Parts,* vol. CP-3, no. 3, p. 119, 1956.
4. S. J. Stein and J. Riseman, "Precision and High Temperature Metal Film Resistors," *Proceedings 1956 Electronic Components Symposium,* Washington, D.C.
5. J. J. Bohrer, W. E. Hauth, Jr., and S. J. Stein, "Metal Film Power Resistors," *Proceedings 1956 Electronic Components Symposium,* Washington, D.C.
6. C. Wellard and B. Solow, "Film Type Precision Resistors," 1956 WESCON.
7. C. Wellard and S. J. Stein, "Molded Metal Film Resistors," 1958 WESCON, San Francisco, California.

8. J. Riseman, "The Electrical Properties of Thin Metal Films," *Transactions of the New York Academy of Sciences,* ser. 11, vol. 19, no. 6, p. 503, April 1957.
9. H. K. Henisch, ed., Conference on Semi-Conducting Materials. N. Mostovetch and B. Vodar, "Electrical Conductivity of Very Thin Metallic Film Evaporated in High Vacuum," Academic Press, Inc., 1951.
10. F. Zandman, S. S. Redner, and D. Post, "Photostatic Coating Analysis in Thermal Fields," *Experimental Mechanics,* Sept. 1963.

Appendix II

Quality Assurance

The ideal electronic component is one that maintains the manufacturer's guaranteed performance characteristics regardless of the conditions of use. It must be free from any change in performance or defects throughout a period of time that far exceeds the lifetime of the circuit or electronic system in which it will be used. Obviously this represents an ideal situation. The user or customer will rightly demand, however, that the characteristics of components supplied by the manufacturer remain compatible with the overall estimated lifetime of the circuit in a normal environment. Military applications, for example, require that the components used in weapons or transmissions systems are able to function continuously at their nominal levels of operation for thousands of hours in a hostile environment (temperatures of between 50 and +125°C, humidity, salt mist, ionizing radiation, intense vibrations, etc.).

The manufacturer must therefore guarantee that a component will operate correctly under the conditions stipulated by the customer. To achieve the desired result, the manufacturing process incorporates a number of steps designed to obtain high performance and satisfactory reliability while minimizing costs. These steps are summarized below.

1. *Performance of the component family.* Proper performance implies that the component is able to operate within the stated limits of tolerance (precision, TCR, noise, etc.) as a function of time, temperature, humidity, and vibration, and avoid catastrophic failures.

2. *Determination and objectives of quality testing.* Tests are designed to provoke potential defects, study their causes, establish relevant correlations, specify the conditions under which quality control should take samples and how it will measure significant parameters (choice of a sampling plan), and determine the relation between the level of reliability of the component and the conditions of operation (nominal, overload, environmental conditions, etc.).

3. *Realistic selection of components for testing.* The selection of components is extremely critical and depends on whether they will be chosen from laboratory specimens (in this case the results of any tests will not be representative of production conditions but may be helpful for making decisions in the future), normal production lines, pilot lines, or after burn-in.

4. *Realistic selection of test conditions.* This is related to the previous step. Laboratory testing of components is designed to trigger the various failure mechanisms as a function of the applied constraints, and determine the eventual operating limits for the part. Repetitive tests performed on components taken from production lines are designed to determine if there are any systematic or unforeseen variations in production conditions. These tests are designed to minimize production costs while maximizing the detection of random variations that might affect production quality.

5. *Accelerated aging.* Because component lifetime is measured in tens of thousands of hours of operation under normal conditions of use, it is essential that testing conditions be scheduled in such a way that the same effects can be obtained in tens or hundreds of hours of operation. Changes in lifetime, for example, can be monitored at temperatures that far exceed those of normal operation. Through the application of the Arrhenius rate law (see pages 284–289), we can calculate the actual lifetime of the batch from which the component was sampled. The danger here is that the tests may trigger component degradation mechanisms whose probability of occurrence under normal conditions is practically zero. The results of laboratory testing are very useful in preventing this from occurring.

6. *Interpreting measurement error.* A systematic study of measurement error should be conducted to avoid rejection (rejection of a component that operates correctly) or acceptance (acceptance of a defective component) errors. This applies not only to the measurement of variations in component performance but especially in estimating the magnitude of the applied stress.

7. *Statistical laws.* Statistical models for describing any variation in performance should be selected on the basis of the failure mechanisms being investigated. Appendix III describes several of these statistical models.

Appendix III

Model for Accelerated Lifetime Testing

A statistical model for accelerated lifetime testing includes the following elements:

(a) A distribution curve for lifetime as a function of time, representing the dispersion among the sampled components.
(b) A mathematical equation relating component lifetime to the magnitude of the applied stress.

Typically the mean value and variance of lifetime distribution are expressed as a function of the stress or acceleration parameter: temperature, pressure, humidity, etc.

The most common distributions used are exponential, normal, Gaussian, log-normal, Weibull, and Arrhenius.

A.1. EXPONENTIAL DISTRIBUTION[1]

A cumulative probability distribution function $F(t)$ represents that fraction of the population to which we apply one or more constraints of known

[1]There are a large number of books on applied statistics covering the different types of distribution described here. Numerical tables for these equations and their specific application to quality control can be found in Juran, J. M. *et al. Quality Control Handbook.* McGraw-Hill, 1974.

origin and magnitude, and which fails in time t. This population has the following mathematical properties:

1. It is a continuous function for all values of $t > 0$.
2. $\lim F(t) = 0$ for $t => -\infty$ and $\lim F(t) = 1$ for $t => \infty$
3. $F(t) \leq F(t')$ for all values of $t < t'$.

The exponential distribution function is given by the following formula:

$$F(t) = 1 - \exp(-t/\theta), \qquad t \geq 0 \tag{A.1}$$

in which θ is the mean time to failure (MTTF). θ is expressed in units of time, for example, hours, months, years, cycles, etc. We can also define the ratio $\lambda = 1/\theta$ as the mean rate of failure.[2] The relation between MTTF and failure rate is valid only for an exponential distribution. The probability density is equal to:

$$f(t) = \frac{dF(t)}{dt} = \lambda \exp(-\lambda t), \qquad (t \geq 0) \tag{A.2}$$

This function corresponds to the lifetime histogram of elements of the population measured over a period of time. Using these equations, we can determine the following fundamental statistical properties:

Reliability function. This represents the remaining population at time t.

$$R(t) = 1 - F(t) = \exp(-t/\theta) \tag{A.3}$$

Reliability is the complement of the cumulative distribution function.

Variance and standard deviation. The second moment, or variance, is equal to θ^2 and the standard deviation is θ. Standard deviation provides a value for the "dispersion" of the failures recorded over a period of time. This dispersion can be determined in practice from a histogram of the lifetime of the elements tested. The exponential distribution can be recognized by the fact that the standard deviation is equal to the mean. This is generally not true for the other distributions described below.

Instantaneous failure rate. Within an interval of time t and $t + \Delta t$, the failure rate of the population surviving at time t is given by the following equation:

$$h(t) = \frac{f(t)}{1 - R(t)} = \lambda \tag{A.4}$$

[2]This equation is valid only for an exponential distribution. Although primarily theoretical, the exponential distribution law can be used in analyzing component failure.

This represents the instantaneous failure rate at time t. This function is important because it can be used to estimate if component failure increases with age or remains stationary.

This explains why θ is referred to as the failure rate. Experimentally we can identify an exponential distribution by the fact that θ does not vary with time. This is not the case for the other types of distribution used for reliability testing.

We also make use of the cumulative random event function:

$$H(t) = \int_0^t h(u)du$$

The function is used experimentally to determine the distribution parameters. For this distribution, $H(t) = \lambda t,\ t \geq 0$. For all distributions, $H(t) = -ln\,[R(t)]$ or $R(t)\ \exp[-H(t)]$.

A.2. NORMAL AND LOG-NORMAL DISTRIBUTIONS

The cumulative normal distribution function is expressed by the following equation:

$$F(t) = \int_{-\infty}^{t} (2\pi\sigma^2)^{-1/2} \cdot \exp(-1/2[(t-\mu)/\sigma]^2)dt$$
$$= \phi[(t-\mu)/\sigma] \quad (-\infty < t < +\infty) \qquad \text{(A.5)}$$

and

$$f(t) = (2\pi\sigma^2)^{-1/2} \cdot \exp(-1/2[(t-\mu)/\sigma]^2)dt$$
$$(-\infty < t < +\infty) \qquad \text{(A.6)}$$

in which μ is the mean time (this parameter is still called the mean time before failure, or MTBF) and σ the standard deviation.

This function is one of the basic distribution functions used in quality control. It has the advantage of being easy to use and interpret but is not always accurate, especially for sparsely populated lots. It is often the source of considerable error when used without caution.

There is another very common distribution used to represent lifetime curves for insulating materials, components (semiconductors, diodes, resistors, etc.), and dielectrics. This is the log-normal distribution, in which:

$$F(t) = \phi((\log(t) - \log(\mu))/\log(\sigma)) \quad (t > 0) \qquad \text{(A.7)}^3$$

The probability density is equal to:

$$f(t) = [0.443/((2\pi)^{1/2}\tau \log(\sigma))] \exp[-(\log(t) - \log(\mu))^2/2\sigma^2] \qquad \text{(A.8)}$$

In which $0.4343 \cong 1/ln(10)$.

The hundredth percentile Z_p of the distribution is given by the following equation:

$$t_p = \text{antilog}[\log(\mu) + Z_p \cdot \log(\sigma)] \qquad \text{(A.9)}$$

In use it is similar to the normal distribution function. For example, assume we have a group of resistors with an average lifetime of 80,000 hours, or $\log(\mu) = 4.90$ (80,000 hours), and a standard deviation of 100 hours, or $\log(\sigma) = 2$.

The portion of the population that fails during the first 8,000 hours will be:

$$F(8000) = \phi((\log(8000) - 4.90)/2) = \phi(-0.50) = 0.3121$$

This means that 31% of the resistors will have failed and there is no point in using these components in applications whose operating life will exceed 8,000 hours.

Figures 1, 2, and 3 represent the shape of the curves for $F(t)$, $f(t)$, and $h(t)$ for the log-normal distribution law.

We see in Fig. 3 that for $\log(\sigma) > 0.5$, $h(t)$ increases suddenly (infant mortality) and then decreases slowly. Because of this the distribution is useful for representing the lifetime of electronic components. It must be used with some caution, however. According to theory, the distribution is 0 for $t = 0$ and then tends toward 0 for $t \Rightarrow \infty$, which is not consistent with experience. It is primarily used in the interval value 0 to 2.5 of function ϕ.

A.3. THE WEIBULL DISTRIBUTION[4]

The Weibull distribution is extremely useful for modeling component lifetimes since the failure rate increases or decreases uniformly. According to various authors it can be used to model situations in which the failure of

[3]The distribution mean is taken as the base-10 log of the mean lifetime. Standard deviation is the base-10 log of the standard deviation of the lifetime.

[4]Weibull, W., "A Statistical Theory of the Strength of Materials." *Ing. Vetenskaps Akad. Handl.* 151 (1939): 15 ff. Juran, J. P., *op. cit.*

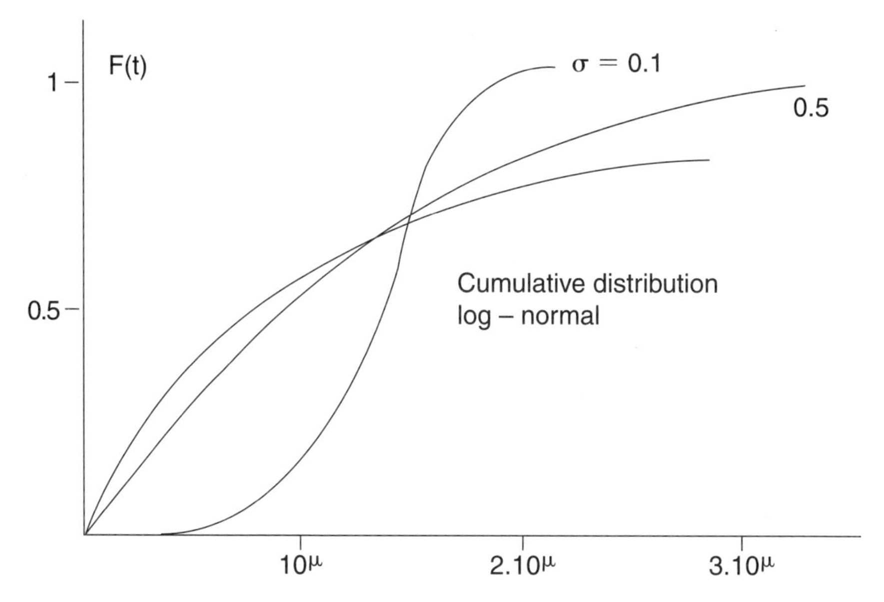

Fig. 1 Representations of $F(t)$, $f(t)$, and $h(t)$ of log-normal distribution for three different values of σ.

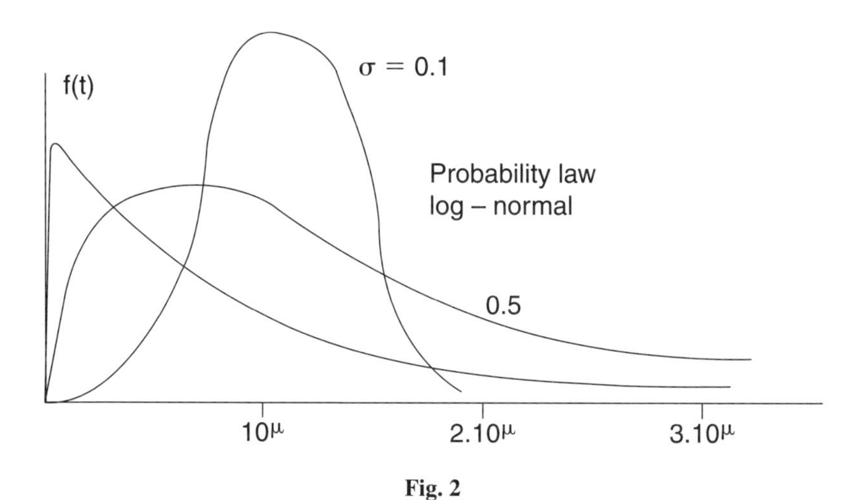

Fig. 2

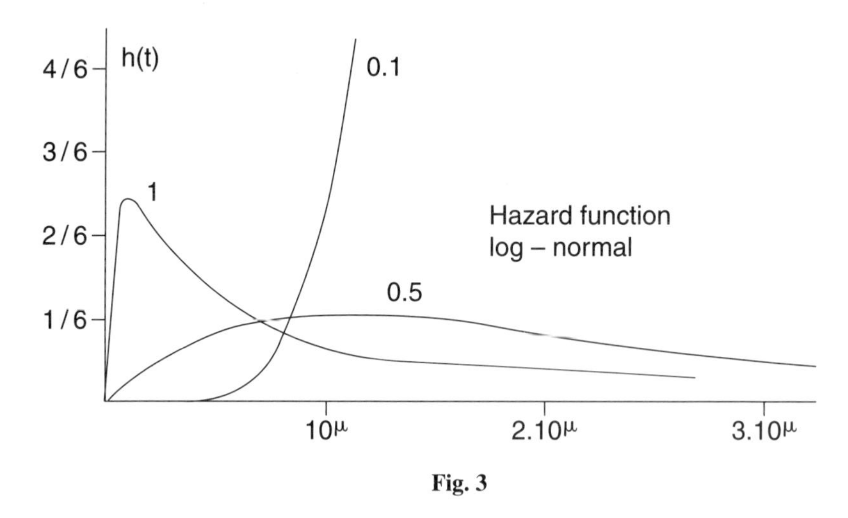

Fig. 3

the weakest link in the component leads to its total failure. Instead of a constant failure rate, as with an exponential distribution, a large number of failure conditions can be created. For example, the sudden failure (open circuit) of a resistor can occur through rupture of the solder on the output leads or overheating of an element, etc. If reliability data can be expressed in the form of "bathtub" curves for component lifetime (infant mortality and end-of-life wear), the Weibull functions can be adapted to individual sections of the curve.

Since the function depends on two parameters, α and β, it can be used to model the curve. The failure rates in these regions can be written, for the general case, as:

$$h(t) = \frac{\beta t^{\beta - 1}}{\alpha^{\beta}} \tag{A.10}$$

where

$\beta < 1$ in the region of early failure

$\beta = 1$ in the region of useful life and constant failure rate

$\beta > 1$ in the wear-out period

The distribution function for the failures is calculated from:

$$F(t) = 1 - \exp\left[-\int_0^t \lambda(u)du\right] = 1 - \exp[-(t/\alpha)^\beta], \quad (t > 0) \tag{A.11}$$

The reliability function is thus $R(t) = \exp[-(t/\alpha)^\beta]$. The hundredth percentile, P, is equal to $t_p = \alpha[-\ln(1-P)]^{1/\beta}$. We see that $t_{0.632} \cong \alpha$ for any Weibull distribution, where α is expressed in units of time (seconds, hours, etc.).

The shape parameter β and scale, or characteristic, parameter α are both positive. β is a dimensionless number whose value is generally between 0.5 and 5.

The probability density is given by:

$$f(t) = (\beta/\alpha^\beta)t^{\beta-1}\exp[-(t/\alpha)^\beta] \tag{A.12}$$

For $\beta = 1$, the Weibull function reduces to an exponential. The failure rate decreases for $\beta < 1$ and increases for $\beta > 1$, as predicted by the tub curve.

If the instants of failure follow a Weibull law, it is easy to determine the characteristic parameters for the distribution. By changing the variable $\tilde{y}_i = Ln[-Ln(1 - \tilde{F}(t_i))]$, we can write:

$$\tilde{y}_i = \beta \cdot Ln(t_i) - Ln\,\alpha \tag{A.13}$$

The values t_i are the instants of failure and $F(t_i)$ is the corresponding cumulative percentage. A least-squares adjustment can be used to determine $\tilde{\beta}$ and $L\tilde{n}\,\alpha$, which are estimates of β and α. These values can also be determined by plotting the data on a Weibull scale.

The principal properties of the Weibull function are given below.

For example, assume that the lifetime of a lot of resistors can be represented by $\alpha = 80{,}000$ hours and $\beta = 0.8$. These values indicate that we are in the infant mortality region [of the curve]. The percentage of failures after a year of operation for the population under consideration will then be:

$$F(8760) = 1 - \exp[-(8760/80{,}000)^{0.8}] = 16\% \tag{A.14}$$

and the instantaneous failure rate per hour $h(8760) = 1.55 \cdot 10^{-5}$/hour. On this basis the lifetime of 80,000 hours is barely sufficient to ensure reasonable reliability. The engineering department will have to improve the value of β, if α cannot be changed, or reciprocally.

Figure 4 represents $f(t)$ and $F(t)$ for the Weibull distribution. We see from the curves that for $\beta = 1$, we again have an exponential distribution.

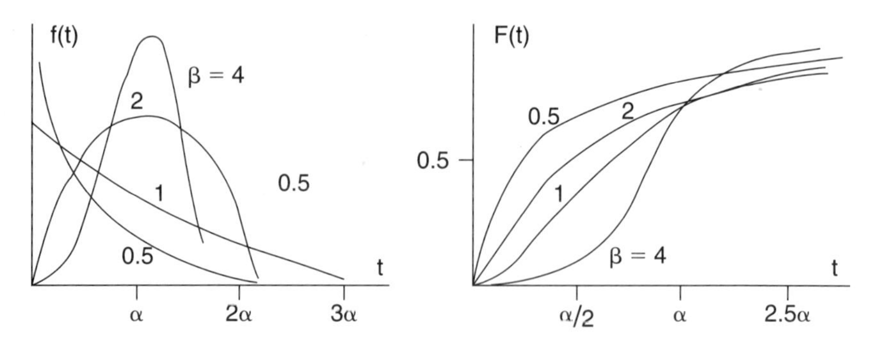

Fig. 4 Representations of $f(i)$ and $F(t)$ of Weibull distribution of different values of coefficient B.

A.4. ACCELERATED AGING AND THE ARRHENIUS MODEL

Techniques of accelerated aging are used to analyze the change in performance shown by electronic components when subjected to stresses that exceed those of normal operation over a period of time. The advantage of this procedure is that it can be used to trigger the degradation process very early in the lifecycle of the component. The results can then be analyzed and we can extrapolate the time over which the component can operate without appreciable loss of performance under normal use. For these methods to be effective, however, the aging model must be adapted to the type of component being tested.

When speaking of resistors, a failure can be said to occur whenever performance degrades beyond a predefined value. For example, when the deviation of the resistance exceeds a certain percentage, say 1% or 3% depending on the type of resistor, or the TCR drifts by more than 50 ppm/°C.

Models used to describe component failure should satisfy a specific number of requirements:

(a) The failure or degradation in performance should not be reversible.

(b) Ordinarily a model is applied to a single degradation mechanism. If there are several mechanisms acting simultaneously, each must be studied using the appropriate model.

The Arrhenius rate law (see Chap. 3) and its variants lend themselves well to the process of resistor aging as a function of temperature. The law is based on the following assumptions:

1. For a given time and temperature or a given voltage, the performance, defined by a measurement result m, follows a log-normal distribution (base 10). In other words, the log of the lifetime of the component follows a normal distribution (m can be the value $\Delta R/R$ of a resistor or ΔTCR/TCR, etc.).
2. The standard deviation σ is constant. Here, σ is not dependent, or is only weakly dependent, on temperature or time. Through the appropriate modifications, however, we can extend the analysis to cases where σ varies.
3. The mathematical relation between the mean m of the log-normal distribution (equal to the base-10 logarithm of the median of the Gaussian distribution), temperature, and time is determined by the following equation:

$$m(t, T°K) = \log(\mu) = A - t \cdot B \cdot \exp(-E/kT) \tag{A.15}$$

The parameters A, B, σ, and E are characteristic of the product, its mechanism of degradation, and the test methods. E is the activation energy, which we discussed in Chap. 3. The equation is based on the assumption that the phenomenon of degradation is controlled by a first-order physico-chemical reaction.

Figure 5 represents the degradation curves obtained using Eq. A.15 plotted on a semi-logarithmic scale. Here, A is the log of the mean m_0 of m at $t = 0$. Equation A.15 can be rewritten as:

$$\log(\mu(t)/\mu_0) = -t \cdot B \cdot \exp(-E/kT) \tag{A.16}$$

This equation provides the percentage log (%) of components at time t that maintain their initial performance rating μ_0.

The percentile of the log of performance $y_p(t, T) = \log[\mu_p(t, T)]$ at temperature $T°$ and time t is given by:

$$y_P(t,T) = m(t,T) + z_p\sigma = A - t \cdot B\exp(-E/kT) + z_P\sigma \tag{A.17}$$

The lifetime distribution can be estimated from Eq. A.15. We assume the component is failing when the log of its performance falls below the value $y_F = \log(\mu_F)$. The percentage of the population $F(t)$ that has failed at time t is the shaded part of the log-normal distribution, expressed as:

$$F(t, T_F) = \phi[(y_F - A + B \cdot t \cdot \exp(-E/kT))/\sigma] \tag{A.18}$$

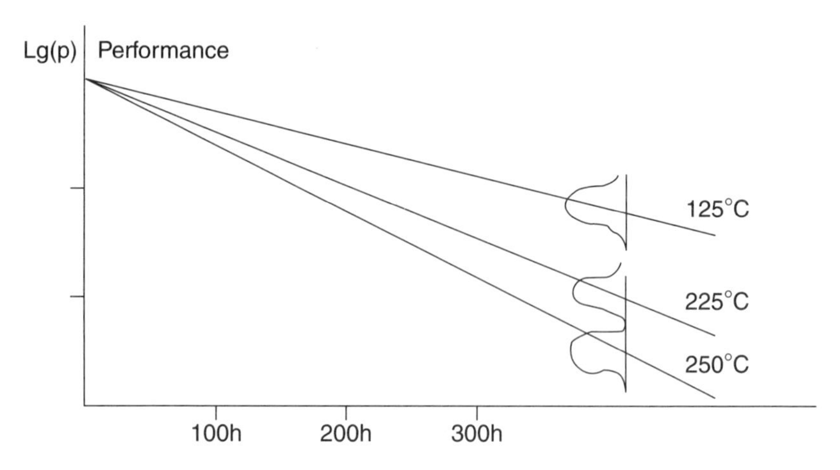

Fig. 5 Degradation curves in function of time and different values of temperature.

Here ϕ is the normal distribution function. This means that the distribution of the probability of failure as a function of time follows a normal distribution whose mean and standard deviation are:

$$M_F = [(A - y_F)/B] \exp(E/kT_F), \quad \sigma_F = (\sigma/B) \exp(E/kT_F) \tag{A.19}$$

Figure 6 shows the relation between the performance degradation curve of a product (degradation of operating voltage at temperature T_F) and the distribution of the probability of degradation over time.

As an example, consider a family of thick-film cermet resistors for which the nominal deviation as a function of temperature and time follows the values shown in Table 1.[5]

The standard deviation for each measurement point on the aging curve is approximately 6% of the mean.

We will now turn our attention to the model represented by Eq. A.15. In this case we can define a loss-of-performance function:

$$P(\Delta R/R) = (1 - \Delta R/R) \times 1 \cdot 10^6$$

We can apply this function to the log-normal law for positive values. Under these conditions the performance at $t = 0$ is $P = 10^6$ and

$$A = \log(P) = 6$$

[5]Sfernice study. Personal correspondence from its research department.

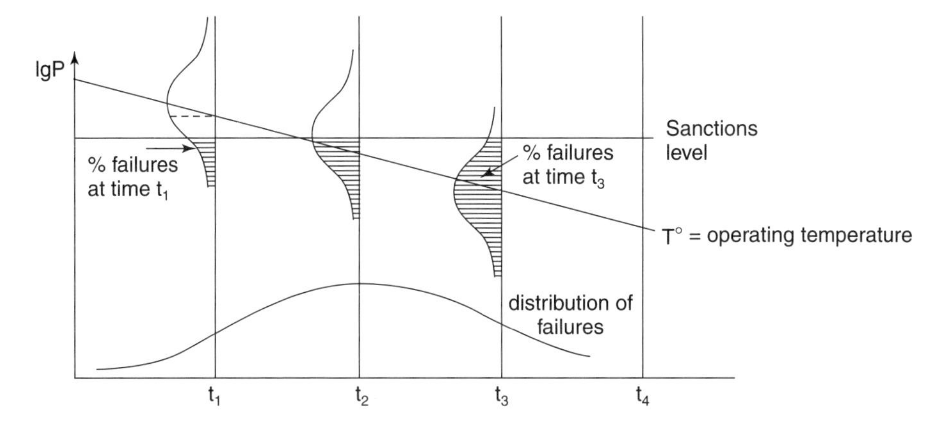

Fig. 6 Relation between degradation values and sanction level for measurements in function of time for a single value T°K.

Table 1 shows that at 1000 hours and 493.16°K, $P = (1 - 0.09) \cdot 10^6$, from which we can determine that

$$m_1(1000{,}493.16) = \log(P_1) = 5.959041392$$

and

$$m_2(1000{,}398.16) = \log(P_2) = 5.996511672$$

It is easy to calculate E by means of the following equation:

$$E = [T_1T_2/(T_1 - T_2)] \ln[(A - m_1)/(A - m_2)] \text{ in °K} \tag{A.20}$$

Table 1 Example of Nominal Deviations with Time at Two Temperature Levels

t (hours)	Deviation $\Delta R/R$ (ppm)	Operating temperature (°C)
5	1000	125
9	1500	125
80	4000	125
100	4300	125
550	6000	125
1000	8000	125
5	6000	220
10	10,000	220
100	15,000	220
500	60,000	220
1000	90,000	220

and

$$B = (1/t)\exp(E/T_1)A/m_1 \tag{A.21}$$

This gives $E = 5090.8°K$ and $B = 1.2459$.

The standard deviation σ is calculated from the log-normal law for P. We need to verify that it is approximately constant for all the measurement points.[6] For a given temperature and time period, the range of the log-normal scale is: $\log(P_{max}) - \log(P_{min})$. Since the law is considered Gaussian, we can assume that the standard deviation will be equal to the sixth [percentile] of the full scale. For our current example, this means that:

$$\sigma \cong 0.00434 \pm 0.0015 \tag{A.22}$$

The loss of performance for this family of resistors is then:

$$A(t, T) = 6 - 1.2459t \cdot \exp(-5090.8/T) \tag{A.23}$$

By applying Eq. A.19 we now have all the information needed to estimate the lifespan of a component population.

By applying a cutoff of 2% (components that deviate more than 2% from the nominal value are considered to have failed) at an operating temperature of 70°C, we obtain:

$$M_F = (0.00877/1.2459) \cdot \exp(5090.8/343.16) = 19{,}512 \text{ hours}$$

$$\sigma_F = (0.003/1.2456) \cdot \exp(5090.8/343.16) = 6{,}676 \text{ hours}$$

Lifetime estimates based on the Arrhenius equations show that, for the given cutoff, the technology or materials used in fabricating this family of components are inadequate. Equation A.16 shows that at time $t = 0 + 100$ hours of operation in an electronic circuit, 0.24% of the components will fail, reaching 50% after 1912 hours of operation.

The model has to be modified slightly to:

(a) increase the activation energy of the failure mechanism, and,
(b) reduce dispersion in the deviation of the aging curve.

After selecting new bake conditions for the resistive ink, measurements show that the activation energy remains unchanged. However, the dispersion σ of the log-normal law is reduced to 0.003–0.001.

[6] In practice the statistical law is sufficiently robust to support significant variations from standard deviation (approximately 100%).

Under these conditions $\sigma_F = 2276$ hours. Using the given cutoff, this results in a failure-free lifetime ($M_F - 3\sigma_F$) of 12,760 hours, which is still not very good. Further calculations show that, based on experimental results, and using a cutoff of 5%, $\mu_F \cong 50{,}000$ hours. The estimated lifetime is 43,000 hours.

In light of the above results, the manufacturer would recommend that these components be used in circuits where high precision and reliability are not required.

Another illustration of the method of accelerated degradation is given by research on another family of chip resistors based on high-precision thin-layer technology. Aging measurements give the following results:

$$\Delta R/R = 2000.0 \text{ ppm at } 220°\text{C and } 1000 \text{ hours}$$

$$\Delta R/R = 200.0 \text{ ppm at } 125°\text{C and } 1000 \text{ hours}$$

The scattering for each aging point has a range of 200 ppm. By modifying Eq. A.15 for this family of components, we obtain:[7]

$$E = 5600\ °\text{K}, \quad B = 7.42 \cdot 10^{-2}, \quad \sigma = 1.4479 \cdot 10^{-5} \qquad \text{(A.24)}$$

that is,

$$\mu(t, T) = 6 - 7.4210^{-2} \cdot t \cdot \exp(-5600.39/T) \qquad \text{(A.25)}$$

Under these conditions, applying a cutoff of 0.1% (1000 ppm) deviation from the nominal value at an operating temperature of 125°C (temperature required for military application), we obtain:

$$M_F = (4.345 \cdot 10^{-4}/7.42 \cdot 10^{-2}) \cdot \exp(5600/398.16) = 7513 \text{ h} \qquad \text{(A.26)}$$

$$\sigma_F = (1.4479 \cdot 10^{-5}/7.42 \cdot 10^{-2}) \cdot \exp(5600/398.16) = 250 \text{ h} \qquad \text{(A.27)}$$

We can therefore estimate a failure-free lifetime of $7513 - 3\sigma_F = 6800$ hours.

At $T = 70°$C the lifetime will be 64,400 hours. These values can be nearly doubled if a cutoff of 2000 ppm is used.

It should be pointed out that the variation in failure-free lifetime is proportional to $[(A - \log(P_F))/\sigma] - 3$ and independent of temperature. This illustrates the well-known fact that for a given cutoff, the scattering of performance data points "controls" the failure-free lifetime of the component.

[7]Vishay–Sfernice study. Personal correspondence.

A.5. OTHER DEGRADATION MODELS

Other degradation models can be used depending on the nature of the external conditions. Some aging processes can be represented by the following equation:

$$\mu(t, C) = A - B \cdot T \cdot C^{\gamma} \quad \text{(A.28)}$$

in which C represents the constraint and γ a coefficient that is characteristic of the product.

Certain aging mechanisms based on the Arrhenius equation are dependent on temperature as well as on an additional constraint C. Under these circumstances, the Eyring equation can be applied:

$$\mu(t, T, C) = A - B \cdot T \cdot \exp[-E/T - \delta C - \varepsilon(C/T)] \quad \text{(A.29)}$$

Here, B, E, δ, and ε are constants that are characteristic of the product. These constants can be determined through the use of statistical regression analysis. Interested readers should refer to the specialized literature on the subject.[8]

[8]See, in particular, *IEEE Transactions on Reliability* and LuValle *et al.*, AT&T technical memorandum 52415-880929-01TM. AT&T Bell Labs, Whippany. LuValle *et al.*, *Journal of Statistical Physics.* 52 (1988): 311–330.

Accelerated Testing, Wayne Nelson, John Wiley & Sons, 1990.

Appendix IV

The Resistance Strain Gage

A.1. INTRODUCTION

The resistance strain gage has, in general, the attributes of a resistor and can be viewed as a variable resistor. The strain gage is designed for maximum resistance change due to mechanical strain, and minimum change in response to other variables such as temperature. The foil resistance strain gages share a common ancestry (and many common aspects of foil processing technology) with precision foil resistors.

In the case of the resistors, the foil is bonded to a special substrate material, selected to minimize the TCR of the resistor (foil/substrate) assembly. Since the foil resistance varies with strain as well as with temperature, and the thermal coefficient of expansion (TCE) of the substrate intentionally differs from that of the intimately bonded foil, which introduces mechanical strain with temperature, the two sources of resistance change can be made to very nearly cancel one another, yielding a resistor with an extremely low TCR.

The foil strain gage, on the other hand, is manufactured with a thin, flexible backing for ease in handling and installing on the surface of some structural or mechanical component for the purpose of strain measurement. The strain gage manufacturer has no control over the type of material to which the gage will ultimately be bonded. Strain measurements may be made on materials ranging from, say, fused quartz with a near-zero TCE to a plastic having a TCE of 50 or more PPM/°F (90+ PPM/°C). If no steps were taken to alleviate the problem, the thermally induced resistance change caused by differential expansion would produce sizable errors in strain measurement, and would sometimes be greater than the resistance change from mechanical strains caused by external loads.

Obviously, it is necessary, then, to compensate or correct for the resistance changes of purely thermal origin when making strain measurements under variable temperature conditions. The development of the self-temperature-compensated (STC) strain gage was a major step in minimizing the thermally induced resistance change of strain gages.* By the proper combination of cold-working the foil and subsequent heat treatment, the strain gage manufacturer produces gages which are tailored for use on different materials having a stepped series of expansion coefficients. The processing is designed to minimize the TCR of the foil/backing/test material combination over the temperature range from about −100°F to +300°F (−75°C to +150°C). In strain gage terminology, the temperature-induced resistance change, expressed in strain units, is referred to as "thermal output."

The actual resistance of a strain gage is not a particularly significant factor in its strain-measuring performance. This is because the strain gage is normally interrogated in a Wheatstone bridge circuit, where the circuit output is primarily a function of resistance ratios. Strain gage resistances are held to reasonably close (fractional percent) tolerances for compatibility with instrument systems characterized by limited bridge balance ranges. Poor gage installation technique may shift the resistance outside of the manufacturer's tolerance band. In this respect, it is noteworthy that a strain gage, to measure strain, must always be bonded to a test part and connected to an instrument system by the user. Proper technique in surface preparation, gage handling and bonding, wiring, and gage protection represents a critical factor in successful gage operation. Thus, the strain gage user is necessarily an active participant, along with the gage manufacturer, in determining the performance of the strain gage installation.

A.2. STRAIN GAGE CONSTRUCTION AND CONFIGURATIONS

Figure 1 illustrates a representative single-element strain gage (enlarged for clarity). The gage pattern is fabricated by a photo-etching process, af-

*J. E. Starr of Vishay-Measurement Group was the principal contributor to the development of the STC technology and in general of the foil strain gage.

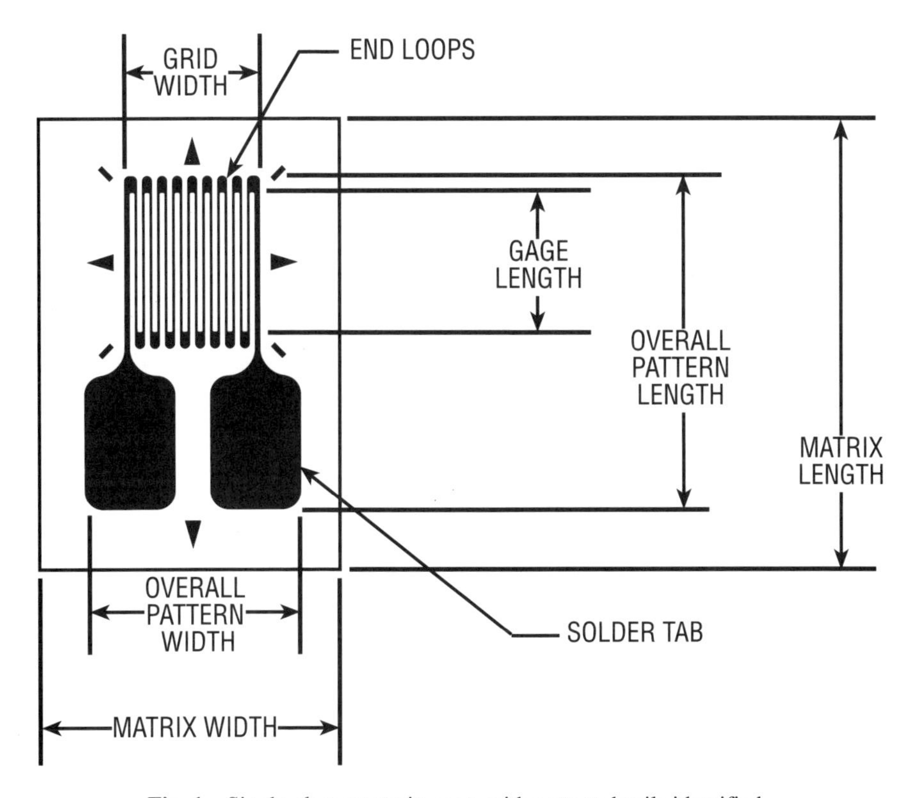

Fig. 1 Single-element strain gage, with pattern details identified.

ter the foil has been bonded to the backing material. The gage grid consists of a series of fine lines, joined in series by larger end loops, and the grid terminates in two generous solder tabs for connecting the gage to electronic instrumentation. As shown in the figure, the pattern also includes alignment marks for accurately locating and orienting the gage on a test surface. The "matrix" designation in the figure refers to the backing material on which the foil is mounted. For strain-measurement purposes, the strain gage is normally connected as one arm of a Wheatstone bridge circuit (Fig. 2). The illustration shows the "three-wire" connection, which places leadwires L1 and L2 in adjacent arms of the bridge circuit. If the leadwires are routed physically adjacent to each other so that they are subjected always to the same temperature, this arrangement results in cancellation of thermally induced resistance changes in the wires. Leadwire L3 carries the bridge output signal with essentially zero current because of the high input

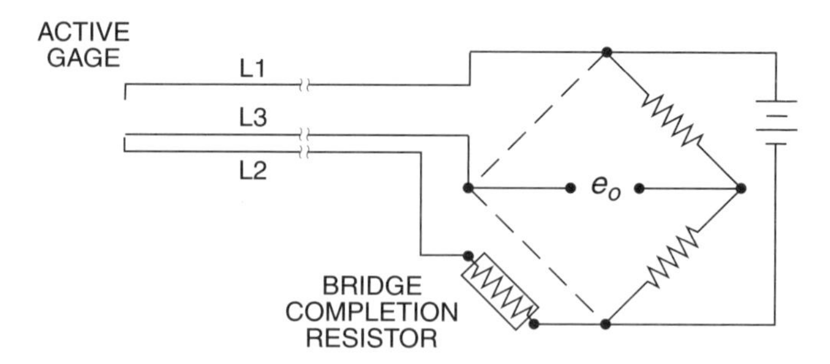

Fig. 2 Strain gage connected as one arm of a Wheatstone bridge circuit. This arrangement is referred to as a "three-wire quarter bridge" circuit.

impedance of the associated amplifier. Strain gage size is specified in terms of "gage length"—the longitudinal span of the straight grid lines. This length defines the "active," or most strain-sensitive, section of the gage. To a stress analyst, the gage length is an important parameter, since the gage tends to integrate, and thus average, the strain under the entire grid, and the stress analyst often wishes to measure the peak local strain at some point on the test object.

A single-element strain gage, as in Fig. 1, is occasionally limited in its (proper) application for stress analysis purposes. This is because most real test objects are characterized by biaxial stress states. In such cases, the maximum and minimum principal stresses at a point are calculated from the principal strains, using the biaxial Hooke's law:

$$\sigma_p = \frac{E}{1 - \mu^2} (\varepsilon_p + \mu \varepsilon_q) \tag{A.1}$$

$$\sigma_q = \frac{E}{1 - \mu^2} (\varepsilon_q + \mu \varepsilon_p) \tag{A.2}$$

Both principal strains must be known to calculate either principal stress, and the directions of the principal strains must be known to measure ε_p and ε_q. It can be shown that in a general biaxial stress state three independent strain measurements at different angles are required to determine the principal strain magnitudes and their directions. To fulfill this need, the strain gage manufacturer supplies a range of strain gage "rosettes" such as that in Fig. 3. The gage elements in a rosette can have any relative orientations, but the most common are: 0°, 45°, 90° ("rectangular") and 0°, 60°, 120° ("delta"). As an example, the principal strains and their directions are obtained from rectangular rosette strain measurements by:

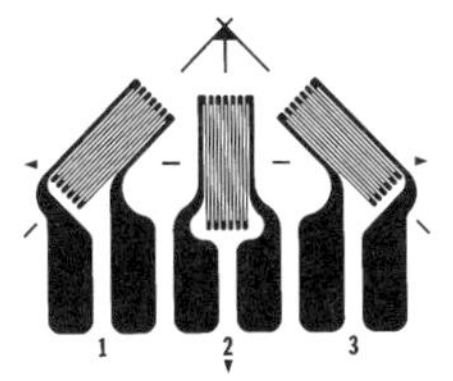

Fig. 3 0°, 45°, 90° rectangular strain gage rosette for determining principal strain magnitudes and directions.

$$\varepsilon_{p,q} = \frac{\varepsilon_1 + \varepsilon_3}{2} \pm \frac{\sqrt{2}}{2}\sqrt{(\varepsilon_1 - \varepsilon_2)^2 + (\varepsilon_2 - \varepsilon_3)^2} \tag{A.3}$$

$$\varphi_{p,q} = \frac{1}{2}\tan^{-1}\left[\frac{(\varepsilon_2 - \varepsilon_3) - (\varepsilon_1 - \varepsilon_2)}{\varepsilon_1 - \varepsilon_3}\right] \pm 90° \tag{A.4}$$

Strain gages are also offered in numerous other configurations for various special purposes. Figure 4 provides a few examples: (a) shear gage for measuring torsion in a shaft; (b) "tee" rosette for use with known principal stress directions; (c) strip gage grouping for determining strain distribution; and (d) special-purpose gage for use on diaphragm pressure transducers.

A.3. SIGNIFICANT CHARACTERISTICS AND PARAMETERS

A strain gage is characterized by the following attributes and parameters: foil alloy and backing material, gage length, resistance, gage factor, transverse sensitivity, thermal output, gage factor variation with temperature, and various user-selected options such as cements and protective coatings, preattached leadwires, encapsulation, and solder dots. The most commonly used foil alloys are constantan (Cu-Ni) and modified Karma (Ni-Cr). Isoelastic alloy (Fe-Cr-Mo), which exhibits a higher gage factor and much greater thermal output, is also employed for dynamic and cyclic applications.

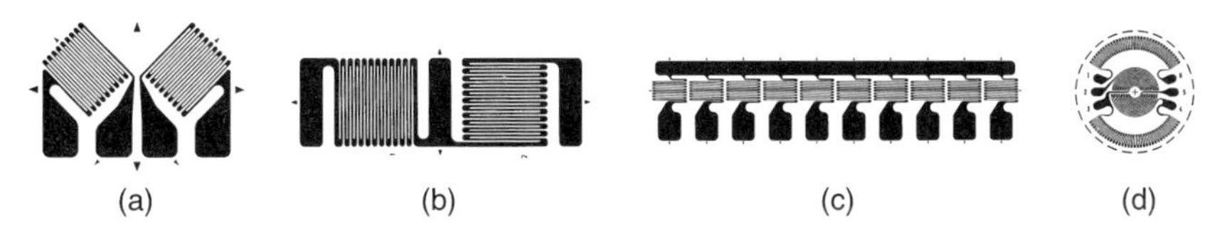

Fig. 4 Special-purpose multiple-grid strain gage pattern (see text).

Backing materials may be polymide or fiberglass-reinforced epoxy. Commercially available gage lengths range from 0.008 in (≈0.2 mm) to 4.0 in (102 mm), with ⅛-in (≈3 mm) and ¼-in (≈6 mm) the most commonly selected lengths. Gages of 1 in (25 mm) or more in length are usually employed for intentionally averaging the strains in inhomogeneous test materials such as concrete.

Standard gage resistances are 60, 120, 175, 350, 500, 1000, 2000, and 4500 ohms, with 120- and 350-ohm gages being the most widely used. Resistance tolerances range from 0.1% to 0.8%, depending on the design and fabrication details associated with a particular gage type. The gage factor of a strain gage is defined as the unit change in resistance divided by the unit surface deformation, or strain. That is:

$$F_G = \frac{\Delta R/R}{\Delta L/L} \tag{A.5}$$

For constantan and Karma alloys, the gage factor is approximately 2.0, and it is about 3.2 for isoelastic alloy and 4.7 for platinum tungsten. The gage factor varies slightly with temperature, but the variation is specified by the strain gage manufacturer to permit user correction for this effect when necessary. As illustrated in Fig. 5, the gage factor of constantan increases with temperature at about 0.5%/100°F (0.9%/100°C), while that of Karma alloy decreases with an increase in temperature. The gage factor variation of Karma is also a function of the self-temperature-compensation (STC) number.

Because of the return loops at both ends of the strain gage grid, and the fact that the individual grid lines are much wider than they are thick, the strain gage is sensitive not only to strains parallel to the grid-line direction, but to transverse strains as well. The ratio of the transverse to the longitudinal uniaxial-strain gage factor, F_t/F_a, is defined as the transverse sensitivity, K_t, of the strain gage. Typically, K_t is positive and less than 1% for constantan gages, and negative and somewhat larger for Karma alloy gages. The transverse sensitivity of each gage type is measured and specified by the gage manufacturer.

When any temperature change occurs during the process of making a strain measurement, the thermal output of the gage must always be taken into account. Thermal output is produced as a result of two additive effects of a temperature change. Since the gage foil has its own TCR, its resistance changes with temperature. But the foil is also strain-sensitive, and because the thermal coefficient of expansion of the strain gage is generally different from that of the test object, the gage is subjected to ther-

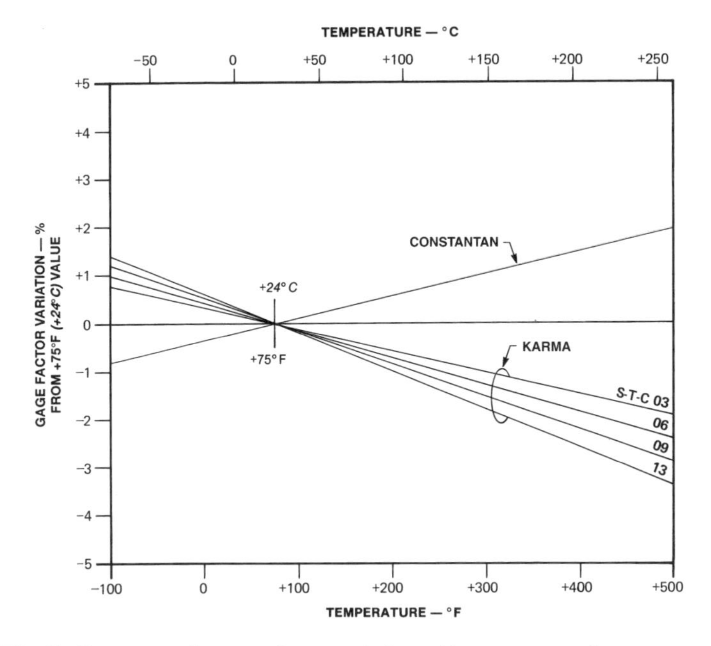

Fig. 5 Representative gage factor variation with temperature for constantan and Karma alloy strain gages.

mal strain with a temperature change. The combined resistance change with temperature, expressed in strain units at a gage factor of 2.0, is the thermal output of the gage. When there is a significant temperature change, the thermal output of the gage can be greater than the strain being measured unless compensation or correction methods are employed. All of Micro-Measurements' (a division of Vishay) constantan and Karma alloy strain gages are self-temperature-compensated. This means that the foil has been cold-worked and subsequently heat-treated so as to minimize the thermal output over the temperature range from about −100°F to +300°F (−73°C to 150°C) when the gage is bonded to the type of test material for which it is intended. Figure 6 is a graph of the thermal output of a typical STC strain gage. STC strain gages arc offered for test materials with thermal expansion coefficients ranging from 0 to 50 PPM/°F (90 PPM/°C). For the best accuracy in strain measurement under varying temperature conditions, the residual uncompensated thermal output shown in Fig. 6 can be corrected for by subtracting the amount in the graph (or given by the included polynomial relationship) from the strain magnitude indicated by the strain-measuring instrument.

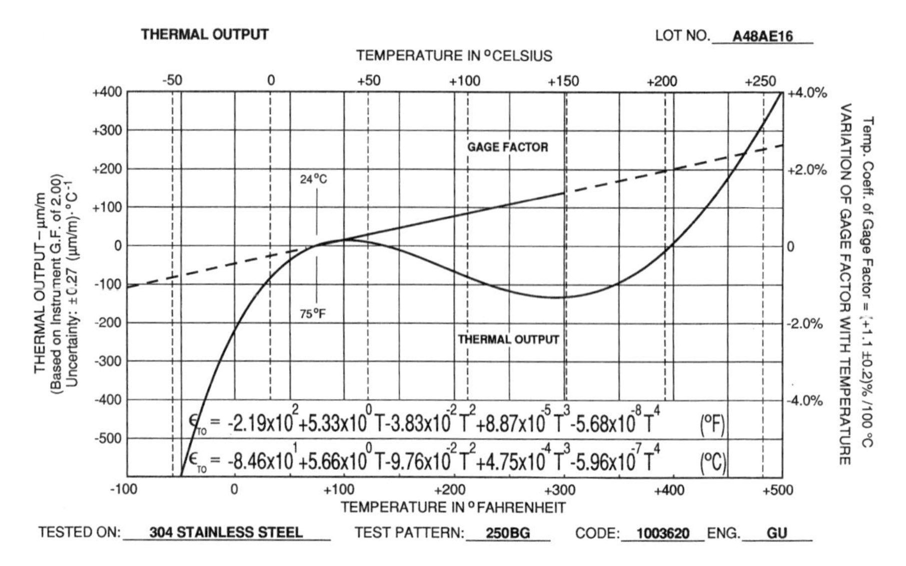

Fig. 6 Graph of thermal output for a typical self-temperature-compensated strain gage, installed on the material for which it is intended.

A.4. CALIBRATION OF STRAIN GAGES

Strain gages are calibrated by the manufacturer for all key operating characteristics such as gage factor, transverse sensitivity, thermal output, and gage factor variation with temperature. These parameters are evaluated with standardized testing procedures specified by ASTM E251 and OIML Recommendation No. 62. In terms of strain gage utilization, all of the foregoing tests are destructive in nature, since the strain gage must be bonded to a metal specimen for the test, and cannot be reused. Because of this, the test procedures always involve statistical sampling to arrive at representative values for the measured variables.

In the case of gage factor, a group of sample gages from each manufactured lot is first bonded along the centerline of a tapered constant-stress cantilever beam, and connected to Wheatstone bridge circuits. Then the beam is deformed to achieve a surface strain of 0.001, or 1,000 μin/in (μm/m), and the amplified output of the bridge circuit is observed to permit calculating the gage factor from:

$$F_G = \frac{\Delta R/R}{\Delta L/L} \tag{A.6}$$

It is noteworthy that the gage factor, as given by Eq. A.5, represents the strain sensitivity of the gage for only a particular state of strain; i.e., corresponding to uniaxial stress in a steel beam with a Poisson's ratio of 0.285. When a strain gage is used under any other conditions than those employed in gage factor calibration, there is always some degree of error due to transverse sensitivity. In other words, any gage which is: (a) installed on a material with a different Poisson's ratio; or, (b) installed on steel, but subjected to other than a uniaxial stress state; or, (c) even installed on steel in a uniaxial stress state, but aligned with other than the maximum principal stress, exhibits a transverse-sensitivity error which may require correction. As a rule, the error is small because the transverse-sensitivity coefficient for most gages is less than 1%. When making rosette strain measurements, however, correction of the indicated strains for transverse sensitivity is usually warranted since at least one of the gage grids is often nearly perpendicular to the maximum principal strain.

Calibration for transverse sensitivity is performed on a different kind of beam than that used for gage factor determination. The beam for evaluating K_t is short, very wide, and stiffened against transverse curvature.[1] As a result of its design, this beam displays almost perfectly uniaxial *strain* over the central 90% of its width. Pairs of strain gages from the same lot are installed on the beam, with one gage aligned with the principal-strain direction, and the other with the zero-strain direction. The outputs of the two gages are then proportional to, respectively, the longitudinal and transverse uniaxial-strain gage factors, F_a and F_t, and their ratio yields the transverse-sensitivity coefficient:

$$K_t = \frac{F_t}{F_a}$$

Strain gage thermal output is measured by bonding the gage to a specimen made from the material for which the gage is processed to be self-temperature-compensated. The stress-free specimen is then subjected to an appropriate range of temperatures while recording the gage output (resistance change) in strain units at a gage factor of 2.0. The result of this test is provided to the gage user in a graph such as Fig. 6, included with each package of gages.

[1]See ASTM E251.

As noted earlier, the gage factor of a strain gage varies slightly with temperature. The error due to this effect is characteristically small, but for greatest accuracy in strain measurement under varying temperatures, a correction should be made. For calibration purposes, the gage is bonded to a short, tapered constant-stress cantilever beam. The beam is placed in a special fixture which permits fixed-displacement formation at several different levels while the assembly is enclosed in an environmental chamber. The beam is then flexed from zero strain to the test strain at different temperatures over the range of interest, typically from −50°F (−45°C) to +500°F (260°C). The variation of the gage factor from its room-temperature value is provided graphically as shown in Fig. 5.

Most strain gage rosettes are etched from the sheet of foil as integral multiple-grid assemblies as shown in Fig. 2. Because the foil has undergone extensive rolling to reduce its thickness to the 100 to 200 μin (2.5 to 5 μm) normally employed in foil strain gages, the foil is no longer completely isotropic in its electromechanical properties. As a result, the rosette grids, which have different angular orientations relative to the rolling direction, require separate measurement of properties such as gage factor.

A.5. APPLICATIONS

Strain gage applications can be divided into two broad categories: (a) strain measurement for purposes of stress analysis, and (b) strain sensing in transducers to measure force, weight, torque, pressure, and other mechanical variables. Although historically, transducer applications motivated the original invention of the resistance strain gage, the needs of the aircraft industry during World War II soon led to the development of the gage as a tool for precision stress analysis. The requirements on the strain gage for stress analysis purposes are different from those for transducer applications. To obtain reliable strain data from which to calculate stresses, the stress analyst requires accurate strain gage properties data, and must commonly adjust the measured strains appropriately to correct for thermal output, transverse sensitivity, etc. The wide range of test articles, materials, testing environments, and other variables encountered in experimental stress analysis necessitates a correspondingly wide range of gage types in terms of gage lengths, grid aspect ratio, number and orientation of grids, self-temperature-compensation, foil and baking types, and details of grid and solder tab design. To satisfy these requirements, the Micro-Measurements division of Vishay offers many thousands of strain gage types, representing all practical combinations of the preceding characteristics.

Although transducer applications of strain gages have always been important, since the resistance strain gage was invented in the late 1930s, this area of gage application has mushroomed following the development of integrated circuits and chip-based microprocessors. Today, transducers having strain gages as the sensing elements are found in numerous consumer and industrial products—particularly in a wide variety of scales. These include, for example, produce scales for supermarkets, bathroom scales, postal and package scales, on-board truck and aircraft scales, etc. Another large area of applications is in pressure transducers for process monitoring and control.

Transducer applications normally involve the use of multiple strain gages in half- or full-bridge circuits as shown in Fig. 7. When gages from the same manufacturing lot are disposed in such circuits, there is automatically complete cancellation of the thermal output, except for the very small gage-to-gage variation. The transducer designer is not usually concerned with the actual gage factor of the strain gage (as long as it is stable and repeatable), since the transducer is normally calibrated for bridge circuit output as a function of the measurement.

The economics of transducer manufacture can become an important consideration in strain gage design. For instance, the process of gage

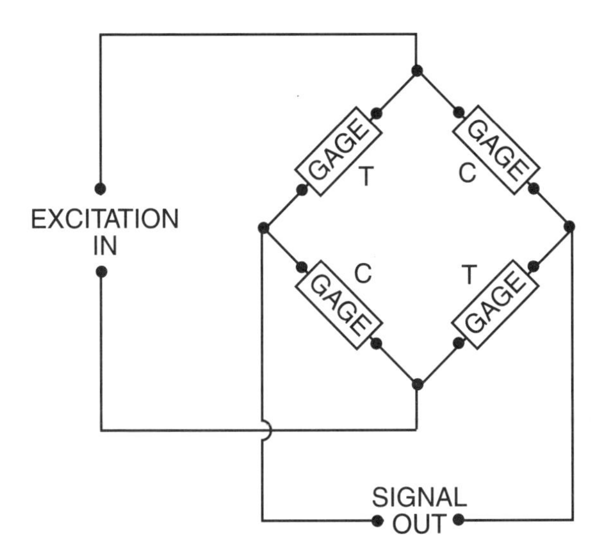

Fig. 7 Wheatstone bridge circuit for a transducer application, with corresponding tension and compression strain gages in pairs of adjacent bridge arms.

installation on the transducer spring element is simplified and accelerated if both gages in a half-bridge arrangement, along with the inter-gage connecting lines, are etched from the foil as an integral assembly. Various other custom design arrangements are employed by different transducer manufacturers, and these become economically feasible because of the high volume of gages used in the large-scale manufacture of a consumer or industrial product.

The typical tolerance on the gage factor of a resistance strain gage is ±0.5 percent. This tolerance forms the absolute limit on accuracy of any strain measurement with the gages. In fact, because of numerous small error sources which may be difficult to fully correct or compensate for, very few stress analysis laboratories can approach the limiting accuracy. The demands for accuracy in transducers are much greater, but these can be met by incorporating compensating circuitry in the design, and by physically calibrating the transducers. Commercially available transducers are offered in different levels of accuracy, at corresponding prices. A low-cost, low-accuracy transducer may be rated at ±1 or 2% accuracy, while a high-precision transducer may have a rating of ±0.01% or better. To achieve such accuracies requires extreme care in the design and manufacture of both the transducer itself and the strain gages employed in it. For example, the spring element of the transducer, to which the strain gages are bonded, may exhibit a very small, but measurable, creep with time under load. The result will be a small upscale drift in the transducer output, and thus an error in the indication of the measurand. One means of compensating for this effect is to shorten the end loops of the strain gages sufficiently to cause a small amount of negative creep in the strain gage itself relative to the spring. When properly done, the two sources of creep will cancel one another to yield a stable transducer output.

Index

D

E

F

G

H

I

J

K

L

M

N

O

P

R

S

T

V

W